REBUILDING AMERICA

REBUILDING
AMERICA Volume 2

Financing Public Works in the 1980s

Roger J. Vaughan

THE COUNCIL OF STATE PLANNING AGENCIES

HALL OF STATES
400 NORTH CAPITOL STREET
WASHINGTON DC 20001

On the cover: *Way to the Citadel* by Paul Klee. Courtesy of The Phillips Collection, Washington, D.C.

© 1983 by the Council of State Planning Agencies
Library of Congress Cataloging in Publication Data

Vaughan, Roger J.
  Financing public works in the 1980s.

  (Rebuilding America; v. 2)
  Bibliography: p.
  Includes index.
    1. United States—Public works—Finance.   I. Title.   II. Series: Vaughan,
Roger J.   Rebuilding America; v. 2.
  HD3885.V36   1983   vol. 2   363'.0973s   83-840
  ISBN 0-934842-21-3   [338.4'3363'0973]

Book and Cover Design by Chuck Myers
Typography by Wordscape
Production by Calvin Kytle Associates

Manufactured in the United States of America

The Council of State Planning Agencies is a membership organization comprised of the planning and policy staff of the nation's governors. Through its Washington office, the Council provides assistance to individual states on a wide spectrum of policy matters. The Council also performs policy and technical research on both state and national issues. The Council was formed in 1966; it became affiliated with the National Governors' Association in 1975.

Funding support for this volume was received in part from the Department of Housing and Urban Development under the Financial Management Capacity Sharing Program. The statements, findings, conclusions, recommendations, and other data contained in this report do not necessarily represent the views or policies of the Council of State Planning Agencies. Reproduction of any part of this volume is permitted for any purpose of the United States Government.

The Council of State Planning Agencies
Hall of States
400 North Capitol Street
Washington, D.C. 20001
(202) 624-5386

Acknowledgments

THIS BOOK WAS PREPARED under a contract from the U.S. Department of Housing and Urban Development. Alan Siegel and Ed Stromberg in HUD's Office of Policy Development and Research provided comments and suggestions throughout this study. This book draws heavily on material prepared under two other HUD-sponsored research projects conducted by the Urban Institute and by the National Conference of State Legislatures. The author is indebted to George Peterson of the Urban Institute and to Ken Kirkland, Steven Gold, and Rick Watson of NCSL for sharing their research.

Many others have assisted by reviewing drafts including Barbara Dyer, Council of State Planning Agencies; John Kamensky, Office of Technology Assessment; Hugh O'Neill, Director, Office of Development Planning, State of New York; Mark Willis, Federal Reserve Bank of New York; Jeff Koshel, senior fellow, National Governors' Association.

Many state officials assisted in responding to requests for information and verifying early drafts, especially for chapter 6. The material presented in chapter 6 is based upon a paper prepared by Klaus Kolb of the Kennedy School of Government. Jeffrey Apfel, Assistant Director of the Office of Development Planning, State of New York, prepared the analysis for, and wrote, chapter 7.

In spite of this assistance, the author remains solely responsible for remaining errors and omissions and for the views expressed, which are not necessarily those of the Department of Housing and Urban Development or the Council of State Planning Agencies.

Foreword

LOCAL GOVERNMENTS face challenging tasks in the 1980s. They have to devise realistic strategies for their cities and mobilize a broad range of resources. They must reexamine the services they provide to establish which are necessary, set priorities for service delivery, determine which services can best be provided by local government directly and which are candidates for alternative service delivery approaches, and actively promote public/private community partnerships.

To help communities address these challenges, the Administration is assisting state and local governments in their efforts to strengthen their management capabilities. HUD's Governmental Capacity Sharing Program is doing this by collecting and disseminating sound ideas and workable practices from the nation's local public and private sectors.

Rebuilding America is part of the Governmental Capacity Sharing Program and is dedicated to encouraging local initiative, self-reliance, and improved performance.

Preface

IN THE LAST FIVE YEARS, the condition of the nation's infrastructure has become a major public policy issue. Reports of the deterioration of roads, bridges, water supply and treatment systems, ports and terminals have documented not only the need to increase spending on repair and maintenance but also the very real danger of physical harm and environmental degradation. Yet, nationwide, the rate of investment in public capital—both new development and maintenance of existing facilities—has declined alarmingly.

As the rate of public capital investment declines, the need for new investments increases. In some areas, the expansion of employment and population has outstripped the capacity of the transportation network and existing water supplies. In cities that have lost population, the configuration of public capital plant is inappropriate for their new service-oriented economic structures. We have also discovered new infrastructure needs: safe and more efficient means of disposing of industrial and municipal waste must be sought; the rapidly growing telecommunications system—a necessary "infrastructure" for new, urban service economies—is encountering severe bottlenecks; soaring energy costs have made resource recovery and recycling necessary developments.

However, the capacity of state and local governments to make these necessary investments is declining. The tax-exempt bond market is no longer a low-cost alternative. Real interest rates are at record peacetime levels. Federal grants have been reduced. State and city governments are wrestling with extraordinarily tight budgets.

There is a yawning gap between state and local fiscal resources and the level of expenditures needed. In *America in Ruins,* Pat Choate and Susan Walter estimate that annual spending on public infrastructure must be quadrupled from the present level of $70 billion. With no increase in federal aid, this would entail a 40 percent increase in all state and local taxes—a prospect that is politically impossible. Innovative financing methods could stretch current public investment dollars by 20 percent—wholly inadequate to meet identified needs. More radical approaches must be sought.

These approaches can include designing cost-effective engineering replacement requirements; cuts in the use of tax-exempt bond revenues to subsidize private development; efficient pricing of public facilities and services; reducing tax abatements and incentives; increased sharing of project costs with private developers; improved planning and management; and reform of financing methods. This book is intended to serve as a guide to state and local policy makers as they seek ways to meet the growing need to rebuild and redevelop deteriorating public works. This volume discusses financing issues, including efficient pricing of public services and facilities, the allocation of responsibility between the public and private sectors, building an efficient partnership between state and local governments, and reducing the cost of debt financing. The companion volume, *Vol. 1: Planning and Managing Public Works Investments in the 1980s*, examines capital planning, budgeting, and management issues.

There are no easy ways to resolve the "infrastructure crisis." Subsidies for business development, which have proliferated in the last decade, will have to be cut. User fees will have to be imposed or increased. New financing mechanisms will have to be designed. Responsibility will have to be shared more equally with the private sector. Some taxes will have to be raised. But we must realize that, in the long run, sustained economic development depends upon adequately financed and efficiently managed public investments, not upon artificially low taxes and temporary subsidies.

Contents

List of Tables and Figures

REBUILDING AMERICA

The "Infrastructure" Question: Meeting Needs or Efficient Investing?

IN 1973, a truck, heavily laden with road-repairing materials, plunged through the Westside Highway in New York City. The surprised driver escaped serious injury, but the event proved a precursor of an issue that now afflicts almost all Americans: the nation's public works are deteriorating.[1] Dams and bridges have been declared unsafe. Roads are impassable. Sewers are backing up. Hazardous wastes are leaking into water supply systems. The rate of public investment in maintaining existing physical infrastructure[2] and in building new facilities has fallen below 2½ percent of GNP—less than half the rate a decade ago. Net investment is close to zero.[3]

State and local officials seem powerless to address the problem. On the one hand, a growing volume of studies indicates that massive increases in public spending are necessary to arrest the rate of deterioration and to develop needed new facilities—perhaps a 300 or 400 percent increase from our present annual spending rate of $70 billion. At the same time, the money to pay for these investments is scarce. Federal aid for capital spending is being cut by 40 percent in real terms between 1981 and 1984. Recession has shrunk state and local tax revenues. Many states and localities face legislative or constitutional limits on debt issues, on revenues, or on both. Tax-exempt interest rates are at record levels. State and local governments must compete in bond markets with huge increases in federal borrowing. Escalating federal budget deficits render any increase in federal aid to states unlikely and make further cuts in aid almost inevitable.

To meet the needs that have been documented in recent studies, state and local governments would have to increase all taxes by about 40 percent. This would be constitutionally or legally impossible in many states, and politically impossible in all. Even if the money were available, it could not be spent efficiently. There are no reliable data concerning the condition of public works because few states, counties, or cities have measured the extent of under-investment, kept records of their capital investment needs, or determined their priorities.[4] In addition, responsibility is split between 100 federal agencies, 50 states, 3,042 counties, 35,000 general purpose local governments, 15,000 school districts, 26,000 special districts, and 209 multistate organizations.

Faced with an apparently insoluble dilemma, local governments have responded in an ad hoc fashion. In California, the aftermath of Proposition 13 has forced cities and counties to make residential developers pay for utility hook-ups. In Michigan, the economic depression in basic industries has forced across-the-board cuts in all public programs. From New York to St. Petersburg, fiscal problems have led to deferred maintenance of public works, which has accelerated the speed of deterioration. To stretch scarce dollars as far as possible, local governments have tried a number of financing gimmicks: in 1981, Massachusetts sold zero-coupon bonds to cut the cost of debt financing; Oakland, California, has sold and leased back its art museum to a private corporation to finance rehabilitation of its Coliseum; and New York City's subway cars will be owned, at least for tax purposes, by a communications corporation.

Innovative financing techniques, however, will not bridge the massive gap between needs and resources. At best, improved bond-marketing techniques, better planning, and innovative management techniques can help current capital spending buy 20 percent more capital investment in real terms. But that is less than 10 percent of the gap. There are no low-cost solutions to span the remaining 90 percent. If we continue with past policies of papering over the widening cracks in our public works, then economic recovery will stumble over ill-paved roads and ruptured water mains. A more comprehensive and radical strategy is required, a strategy that encompasses eight elements:

1. *Redesigning Engineering Replacement Standards and Improving Maintenance Procedures.* Many of the standards that are used by federal, state, and local governments to measure investment needs and to guide maintenance procedures can, with no harm, be modified to delay the replacement cycle and to ensure that what money is spent is allocated to high-priority projects. In an analysis of public investment financing policy, Peterson and Miller conclude:

> Although needs gaps are usually thought to be filled by additional spending, they can be closed by reconsidering and reducing the needs standards that have given rise to the investment gap. This is not a matter of redefining infrastructure needs to make them disappear, but of recognizing that needs always exceed resources and that, with budgets as tight as they now are, priorities must be selected even within "needs" categories (Peterson and Miller, 1981, p. 2).

The issue is not only defining replacement cycles that reflect realistic priorities but also designing engineering standards that are cost effective. For example, more than half the road bridges requiring replacement are on the list because the bridge is narrower than the approach roads, not because they are structurally unsound. Public investment, like private investment, must be guided by a careful comparison of costs and benefits, not by simple engineering algorithms.

2. *Reducing Subsidies to Private Investment.* State and local governments must cut back on using tax revenues and the proceeds from tax-exempt bond sales to finance private sector capital investments. In 1981, more than half the proceeds from long-term bond sales were used to finance projects that are not, traditionally, the responsibility of the public sector, including private hospitals, industrial development projects, pollution control facilities for private companies, convention centers, and university dormitories. State and local governments do not have the resources to maintain and develop public infrastructure—roads, bridges, sewer and waste systems, schools, and parks—and to subsidize investments by private firms. The National Economic Recovery Tax Act of 1981 provided deep incentives for private investments while raising the cost of public borrowing. Using scarce state and local resources to provide further incentives is not fiscally prudent although it is politically difficult for a state to reduce these incentives unilaterally.

3. *Improving Capital Planning, Budgeting, and Management.* A comprehensive public investment strategy will require much more sophisticated planning and budgeting procedures and the design of more effective management techniques for disbursing funds and for operating public facilities. Shepard and Goddard summarize the need for better planning:

> Strategies and programs to restore the infrastructure to its role of providing efficient service to human and economic goals must be preceded by deliberation and the capacity to act—the country has learned all too well that massive doses of money, even if they were available, can contain the disease but cannot provide a complete cure. Time conscious planning and the capacity to analyze appropriate capital development strategies can be as valuable a step as the generation of the dollar resources needed for restoration and revitalization (Shepard and Goddard, *no date*, p. 18).

The failure to develop comprehensive needs assessment, planning, and budgeting techniques is reflected not only in the misallocation of resources among projects and the failure to prevent serious breakdowns, but also in increasing delays in undertaking projects. In *As Time Goes By*, Dr. Pat Choate (1979) documents a backlog of funded but unconstructed federal, state, and local capital projects of $80–$100 billion. With high inflation, these delays have escalated project costs by as much as 12 percent annually or over $1 billion each month. Not only do we need to develop effective planning mechanisms for capital projects, but we need to develop mechanisms to speed up the permitting and contract-letting procedures to avoid costly delays. New administrative structures will be needed to ensure that public facilities are efficiently managed.

4. *Project-Specific Cost Sharing Arrangements with Private Firms.* Some public capital projects are made for the exclusive benefit of major private sector developments. The most obvious examples are new towns constructed to house the labor force for large-scale energy developments in the West. Smaller examples include streets, sidewalks, and utility hook-ups for residential subdivisions, and highway and railroad spurs for new manufacturing facilities. The major beneficiary of these investments should be required to pay at least part of the investment costs.

5. *Charging for Public Services.* Bridging the infrastructure gap will require much greater reliance on "user fees" for public services, such as waste disposal and water supply, and recreation. User fees should not be regarded as merely a source of revenues. Efficient pricing encourages efficient use of services and will reduce the amount of new construction needed. For example, if water use is metered and charged for, supply capacity will not necessarily have to expand in proportion to population or employment growth. Paying for each gallon of water consumed encourages conservation, which means fewer new reservoirs and treatment plants. Less public capital spending will be needed if user fees are more widely applied, where the administrative costs are not prohibitive and equity issues are manageable.

6. *Improving Bond-Financing Mechanisms.* A little more than half of all state and local capital spending is financed through revenues from the sale of tax-exempt bonds—including general obligation and limited obligation issues. The disorganization and decentralization of the bond market in some states have contributed to high financing costs—there are over 1.5 million different tax-exempt issues outstanding. The costs of issuing these bonds can be reduced through state bond banks, state bond guarantees, and state loans to localities. In addition, new revenue sources can be found to service the debt and improve the credit rating.

7. *State Assistance to Local Governments.* Local governments—cities, counties, and special districts—directly invest about twice as much as state governments in capital projects, although many of the funds are provided from federal and state sources. Many local jurisdictions have reached statutory or constitutional bonding limits and are facing extreme fiscal pressures as a result of the economic recession and cuts in federal aid. Some increase in state aid (both dollars and planning assistance) will be necessary for those cities and counties with extensive infrastructure needs and slender fiscal resources. The state can help through technical and planning assistance, intrastate revenue sharing, direct loans, categorical project grants, or state assumption of local responsibilities.

8. *Increasing Public Capital Investment.* The final element in bridging the gap is for increased spending on public works. The revenues will come from user fees, increases in existing tax rates, and perhaps from reducing tax abatement and exemption programs for business. Although these steps will not be politically popular, taxpayers may be less hostile if the revenues from increased tax rates are directly committed to specific purposes and projects.

These eight steps are complementary elements of an effective program of public capital investment. They reduce the apparently overwhelming infrastructure crisis into a series of manageable policy issues. Instead of asking the unanswerable question—How do we finance a quadrupling of state and local infrastructure spending?— we must examine three questions. How should responsibility for infrastructure financing be shared between the public and private sectors? What is an efficient level of public investment in, and maintenance of, capital facilities? And, finally, What is the most efficient way of financing public investments? The goal is to make efficient investment decisions, not to meet ill-defined needs.

This book discusses financing options (steps 4, 5, 6, 7, and 8). In the companion volume, *Planning and Managing Public Works Investments in the 1980s*, planning, budgeting, and management issues are analyzed in detail (steps 1, 2, and 3). It is intended to assist state and local officials—those charged with planning, budgeting, and policy development—to think strategically about public capital investment policies. It reviews the alternative policies that can be pursued in designing user fees, improving the operation of the bond market, sharing responsibility with the private sector, and providing increased assistance to local governments. It is not intended as a technical manual in the arcane procedures related to bond sales, leasing, and negotiated investment strategies.

The following chapter describes current capital spending by federal, state, and local governments and their fiscal capacity to increase the level of spending. The third chapter provides a brief overview of the principles that must guide the design of a public capital investment strategy, including a discussion of the allocation of responsibility between the public and private sectors, and the selection of alternative financing methods. In chapter 4, user fees are discussed. The term *user fees* is applied to all fees, charges, and dedicated tax revenues (where the level of taxes paid is related to the use of a facility, e.g., gasoline tax revenues applied to road repair) that are used to finance the operating and maintenance costs of public works and new construction. The chapter opens with an examination of equity and efficiency issues and concludes with an analysis of opportunities to apply user fees to different types of public facilities.

Debt-financing issues are discussed in chapter 5. The opening sections describe the operation of the tax-exempt bond market and how recent changes in federal policy have influenced the market. The final section explores ways of reducing the cost of debt financing including technical assistance, state guarantees for local issues, bond banks, zero-coupon bonds, and state incentives for local capital investments. The sixth chapter analyzes the special problems that state and local governments face with respect to financing public infrastructure necessary for large-scale private resource development projects. It outlines how costs can be shared between the public sector and private firms, and how states can assist localities in avoiding the adverse effects of large development projects. Chapter 7 outlines how state and local governments can use leasing, service contracts, and the outright sale of public facilities to private firms to reduce the costs of public capital investments. Chapter 8 summarizes the major conclusions from the preceding chapters. It explores state-local fiscal relations and ways to increase state aid to localities for capital investments. It also briefly examines how alternative federal actions may influence the design of state and local capital strategies.

By carefully selecting financing techniques, and through complementary reforms in capital planning and management procedures, the infrastructure problem can be reduced to reasonable and tractable dimensions. A prudent state investment strategy can lay the foundation for stable and sustainable economic growth.

CHAPTER I NOTES

1. See Choate and Walter, 1981; Choate, 1982; Dossani and Steger, 1980.
2. The term *infrastructure* refers to facilities and equipment necessary or desirable for the delivery of services to meet social and economic needs. The most exhaustive classification is provided in a study conducted by Abt Associates for the U.S. Department of Agriculture. They identified thirty-seven categories under two major headings: service facilities and production facilities. These categories are reproduced below (see also Holland, 1972). Most studies focus on water systems, sewer systems, streets and highways, mass transit and bridges (see Dossani and Steger, 1980; U.S. Department of Commerce, 1980; Humphrey et al., 1979).

Categories of Public Infrastructure

I. SERVICE FACILITIES	II. PRODUCTION FACILITIES
Education	*Energy*
1. Elementary Schools	1. Direct Power Suppliers
2. Middle Schools	
3. Secondary Schools	*Fire Safety*
4. Public Libraries	
	1. Fire Stations
Health	2. Vehicles
	3. Communications Systems
1. Hospitals	4. Water Supply and Storage Facilities
2. Nursing Homes	
3. Ambulatory (Outpatient) Care Facilities	*Solid Waste*
4. Ambulatory Dental Care Facilities	1. Collection Facilities and Equipment
5. Ambulatory Mental Health Facilities	2. Disposal Sites

6. Residential Facilities for
    a. orphans and dependent children
    b. the emotionally disturbed
    c. alcoholics and drug abusers
    d. the physically handicapped
    e. mentally retarded
    f. blind and deaf
7. Emergency Vehicle Services

Justice

1. Law Enforcement Facilities
2. Jails

Recreation

1. Community Recreation Facilities

Transportation

1. Railroad Facilities
2. Airports and Related Facilities
3. Streets and Highways (including bridges)
4. Inter- and Intra-Community Transit

Telecommunications

1. Cable Television
2. Over-the-Air Television
3. Disaster Preparedness

Waste Water

1. Sewer Mains and Collection Systems
2. Treatment and Disposal Systems

Water Supply

1. Community Systems
    a. Storage Facilities
    b. Treatment Facilities
    c. Delivery Facilities
2. On-Site Wells and Cisterns

---

3. See U.S. Department of Commerce, 1980, especially appendix A.

4. For a discussion of the data problems, see Abt Associates, 1980; Choate and Walter, 1981; Dossani and Steger, 1980; U.S. Department of Commerce, 1980.

Source: Abt Associates, National Rural Community Facilities Assessment Study: Pilot Phase, Final Report, March, 1980, p. 16. See also appendix A-1 for definitions used by others.

CHAPTER II

Public Capital Investments: Present Patterns and Fiscal Capacity

PUBLIC CAPITAL INVESTMENTS—in everything from schoolhouses to sewer systems—are made by all levels of governments, from federal agencies and state authorities to city and county governments and thousands of special districts. Funding sources are equally diverse and include appropriations from general revenues, intergovernmental grants, revenues from fees and charges, and the proceeds from tax-exempt bond sales. In some cases, private resources are used. The allocation of responsibility among different levels of government for raising the funds to pay for the maintenance and development of public works and for actually managing the process of capital spending varies among different types of projects and among different states. So, too, does the choice of financing mechanisms. Constitutional restrictions, legislation, and historical accident have all contributed to divergence in practice and capability among different governments. As state and local governments develop capital investment strategies, they must operate within these legislative and institutional constraints. What is efficient and feasible in one place may be impossible in another.

The purpose of this chapter is to provide a brief description of present public capital investment practices. It gives a context in which to place the more specific discussions and recommendations in the following chapters. The first section defines what is meant by *public capital investment* and its role in public service delivery. The second section describes current patterns of public investments and how the responsibility for finance and management is allocated among different levels of government. The third section provides an overview of recent fiscal patterns of state and local governments, including changes in revenue sources and expenditures.

Definitions and Concepts

Economic development requires capital investments. We must set aside some of our output of goods and services to expand our ability to produce in the future. Investment not only provides the machinery and resources necessary to generate new jobs for the growing labor force but also increases the productivity of existing workers so that real incomes will grow. Capital accumulation can take many forms. It includes investments in research to develop new products and processes, in new plant and equipment, in maintenance and repair of existing machinery. The acquisition of education and new skills by the work force, the managing and development of natural resources, and the provision of the underlying public structures necessary to provide public services and to sustain development are also forms of capital accumulation.

Neither public nor private capital investments are ends in themselves. Both are necessary to expand production of goods and services demanded by households. Although it has become fashionable to downplay the usefulness of public spending, the difference between public and private capital spending does not arise because of differences in their contributions to economic development. A sound transportation system is as important to our economy as modern steel-producing facilities. The difference is that, while the returns to a private investment accrue in most part to the private investor, the benefits of a public investment accrue to a much broader population and are less easily measured. Traditionally, the responsibility for certain types of investment has been assumed by the public sector because market structures do not enable a private investor to earn a sufficient rate of return, even though the investment would yield benefits in excess of costs. The network of urban streets, for example, is of tremendous value to local residents and businesses alike but cánnot easily yield a stream of revenues to reward a private firm undertaking its construction.

The public sector has also assumed responsibility for certain types of services—and therefore the capital investments they require—for reasons of fairness, or equity. Primary and secondary education, for example, have become a local government responsibility in order to guarantee that all children can learn the basic skills needed in society, regardless of their parents' income and regardless of the admissions policies of local private schools.

Unfortunately, the definition of what constitutes an appropriate public responsibility is vague. Indeed, it changes over time with technological development and evolutions in market structures. It also responds to shifts in public tastes and in public perceptions of problems. Environmental issues became a public responsibility as the costly and hazardous consequences of pollution became apparent. In the past decade, "economic development" has become a public responsibility in some areas because of job loss and persistent high unemployment rates. Public and private responsibilities for capital investments increasingly overlap.[1]

Clarifying public and private roles is a necessary first step in designing a state investment strategy.

It is also necessary to clarify the role of public capital in economic development. Many researchers distinguish between social and economic infrastructure. But this is too narrow a view. A very large share of state and local capital expenditures are for facilities that are concerned primarily with improving the quality of "human capital" which is as vital for economic growth as is physical capital. Education facilities, from primary schools to advanced research centers, are essential to the enhancement of the skill levels of the work force, and most studies have found that investments in these facilities are as important as direct investments in plant and equipment in determining the overall rate of growth of the national economy. Health facilities, from community centers to advanced medical research institutes, have contributed to the steady increase in the health of the work force, another important factor in sustaining economic growth. Human capital investments will prove increasingly necessary as the rate of growth of the work force slows in the next decade and as the technological revolution creates growing demands for highly educated and trained labor. Together, health and education capital expenditures account for nearly half of those public capital investments financed from state and local resources. Although these projects are normally regarded as expenditures—to be cut when the fiscal tide ebbs—they should more properly be regarded as "investments" yielding future benefits in terms of reduced welfare dependency, reduced unemployment insurance expenditures, increased incomes, and more rapid job growth. Infrastructure investments, therefore, refer not only to projects such as roads, bridges, ports, and water supply facilities, but also to the facilities that improve the fitness and skill of the work force.

In spite of the critical importance of state and local capital spending to sustainable economic growth, the economic policy debate in Washington has emphasized private investment incentives almost to the exclusion of public investments, in spite of mounting concern over the condition and level of public investments.[3] In fact, the ability of state and local governments to finance capital investments out of bond revenues has been harmed by recent changes in the federal tax code.

In summary, public capital investments are made to meet public demands for public services. They must be evaluated with the same rigor as private investments, even though the measure of the benefits is not as simple. The services provided through these public facilities are a vital factor in ensuring continued economic growth and are necessary to complement private capital investments.

Current Patterns of Public Investment

Analyzing public capital investments is not easy.[2] Neither the federal government nor many states maintain compatible and detailed capital budgets from which data on new construction, maintenance, and repair expenditures can be compiled. However, using the data that are

15

available, we can present an overview of current investment patterns and recent trends. This section reviews
three major aspects of public works investments: (1) the
rate of investment, focusing on the decline during the past
decade; (2) the lack of any clear allocation of fiscal and
management responsibility among federal, state, and local
governments; and (3) changing priorities among different
types of investments.

The Rate of Investment

By almost any measure, the rate of investment in public works is declining. In 1980, federal, state, and local
governments spent $70 billion on nondefense capital proj-
ects—about 2.3 percent of GNP. This is much less than
the 4.8 percent investment rate in 1970, and the peak rate
of over 5 percent in 1965. As a share of total state and local
expenditures, capital investments have declined from 27
percent in 1960 to 15.4 percent in 1980. In constant (1972)
dollars, capital expenditures declined from $53.6 billion
in 1968 to $46.4 billion in 1979 (table 1).

Although the performance over the last decade shows
a sharp decline in the rate of investment, there has been
some recovery in recent years. The lowest rate of investment was reached in 1977, and increased between 1978
and 1980. Peterson and Miller summarize:

> In retrospect, the period 1980–82 may prove to be a water-
> shed for infrastructure financing. Preliminary data point
> to the possibility of a sharp reversal in capital budgets,
> especially in large cities, as the infrastructure problem
> receives more recognition. However, the picture is far from
> clear. . . . The need for infrastructure investment certainly
> has seized public attention, but actual investment growth
> may have been forestalled by high interest rates, federal
> aid reduction, and local budgetary pressures (Peterson and
> Miller, 1981, p. 13).

The ability of state and local governments to sustain
needed capital investment is obviously weakened by the
condition of the economy and by cuts in federal aid. Preliminary data for 1982 indicate a sharp decline in capital
spending in the third quarter. Preliminary data for 1981,

Table 1

Government Expenditures for Fixed Capital Investment in Current and Constant Dollars, Selected Years

($ in billions)

	Current Dollar			Constant 1972 Dollar		
		Expenditures			Expenditures	
	Total	Federal*	State & Local	Total	Federal	State & Local
1960	$26.6	$13.1	$13.5	$40.3	$18.5	$21.8
1965	34.2	14.1	20.1	48.4	18.8	29.6
1968	41.5	13.9	27.6	53.6	16.8	36.8
1970	42.8	14.0	28.8	48.2	15.3	32.9
1971	45.6	15.4	30.2	48.1	16.1	32.0
1972	48.9	18.0	30.9	49.3	18.1	31.2
1973	51.5	17.9	33.6	48.8	17.3	31.5
1974	58.4	17.7	40.7	48.0	15.9	32.1
1975	63.0	21.2	41.8	48.3	17.1	31.2
1976	63.3	23.4	39.9	48.5	17.6	28.2
1977	64.2	25.2	39.0	43.6	17.5	26.1
1978	75.7	29.0	46.7	46.2	18.0	28.2
1979	83.7	32.9	50.8	46.4	18.9	27.5

Notes: Includes federal military expenditures.

The data presented are for investments in structures only and are therefore smaller than the investment totals shown in table 3, which include land costs and equipment, as well as maintenance and repair expenditures.

Source: Bureau of Economic Analysis, U.S. Department of Commerce, unpublished data, adapted in Peterson and Miller, 1981.

compiled by the Bureau of Economic Analysis, show sharp cuts in public works spending. Infrastructure maintenance has proved vulnerable when cuts in spending are made (Peterson, 1979; GAO, 1982; Choate and Walter, 1981). Mayor Helen Boosalis of Lincoln, Nebraska, has said that "in the choice between laying off police or maintaining sewers, sewers always lose" (*Newsweek*, 8/2/82, p. 21). Between 1977 and 1979, the city of Raleigh cut capital spending by 60 percent (Wolman and Davis, 1980). In 1978, San Francisco cut investments by 45 percent and St. Petersburg cut capital spending by 43 percent (Ibid). The Joint Economic Committee (1980) conducted a survey of major cities and concluded that neglect of public capital "appears to be the single greatest problem facing our nation's cities."

The Allocation of Responsibility

Although most actual capital spending is done by local governments, a large share of the funding comes from the federal government. During the last decade, federal aid funding increased from 20 percent of state and local capital spending in 1970 to 40 percent in 1980 (table 2). Federal grants have not only increased the level of investment spending by state and local governments, they have also changed investment priorities as localities adjusted their expenditure patterns to capture money from federal agencies. The lack of any clear allocation of responsibility among federal, state, and local governments for capital investments has led to considerable duplication, and costly mandates, regulations, and requirements.

The increase in federal aid arose from two perceptions. The first was that state and local governments were doing a poor job in meeting certain national economic, social, and environmental objectives. The waste water treatment program, financed through the Environmental Protection Agency, was undertaken because local actions were inadequate to protect and restore the nation's lakes, waterways, and oceans. By the late 1970s, the program had surpassed the interstate highway system as the largest federally funded public works program. The expansion of federal funding of mass transit during the decade was intended to reduce the nation's dependence on imported oil. Similarly, the Community Development Block Grant program was an attempt by Congress to ensure that cities met the needs of their poorer communities.

The second perception was that the fiscal position of state and local governments was worsening as a result of the rapid growth in expenditures on social programs, in addition to job and population loss in central cities, and inflation. Although New York City had the most publicized brush with bankruptcy, many major cities and older states did not fare much better. Revenue sharing and categorical grants-in-aid were distributed according to indicators

Table 2

Federal Capital Aid Compared to Total State-Local Capital Spending, Selected Years

($ in billions)

Year	Capital Grants	Total State-Local Capital Spending	Federal Capital Aid as a Percent of Total Spending
1970	$ 5.9	$28.8	20.5%
1975	9.0	41.8	21.4
1976	11.4	39.9	28.5
1977	16.4	39.0	42.1
1978	18.0	46.7	38.5
1979	19.8	50.8	39.0
1980	22.3	55.9	40.0

Source: Peterson and Miller, 1981, p. 9

measuring need and fiscal capacity. It is still unclear whether the urban fiscal crises in the mid-1970s were due to a reversible overcommitment to public spending or were due to an irreversible decline in the fiscal base. The economies of Massachusetts and New York have recovered both their economic and fiscal health in a surprisingly short time. However, some cities are still awaiting the first signs of an economic renaissance. Federal fiscal assistance allowed localities to spend more than local resources allowed and to avoid unpopular tax increases. But it also muddled lines of responsibility and accountability.

The federal government is responsible for only a small fraction of *direct* capital spending (if expenditures for national defense and international relations are excluded), a little over $8 billion out of a total of $71 billion. Yet it is responsible for financing nearly half of state and local capital investments (table 3). There are some patterns in the allocation of spending responsibility among the three levels of government. The federal government has the major responsibility for investments in natural resources[4] (the acquisition of land for preservation and recreation), for some pollution control facilities, and for water transportation, ports, and terminals (Army Corps of Engineers projects). State governments are primarily responsible for

Table 3
Direct Governmental Capital Outlays in FY 1980, by Function and Level of Government

($ in millions)	Total Capital Outlay	
Function	All Governments	Federal Government
All Functions	99,386	36,492
National Defense and International Relations	28,161	28,161
Space Research and Technology	239	239
Education	10,834	97
Local Schools	7,362	—
Institutions of Higher Education	2,972	—
Other	500	97
Highways	19,265	132
Health and Hospitals	3,116	673
Sewerage	6,272	—
Parks and Recreation	2,320	297
Natural Resources	5,098	4,046
Housing and Urban Renewal	2,565	317
Air Transportation	1,542	151
Water Transport and Terminals	1,626	1,003
Utilities	9,933	—
Water Supply	3,335	—
Electric Power	4,572	—
Transit	1,921	—
Gas Supply	105	—
Other Functions	8,413	1,375

Note: Because of rounding, details may not add to totals. Local government amounts are estimates subject to sampling variations; see—Represents zero or rounds to zero.
See note under table 1 for compatibility of data.

| Total Capital Outlay | | | Construction Outlays | |
| | State and Local Governments | | | |
Total	State	Local	All Governments	Federal Government
62,894	23,325	39,568	58,410	6,918
—	—	—	2,038	2,038
—	—	—	138	138
10,737	3,182	7,555	7,460	80
7,362	323	7,039	5,186	—
2,972	2,458	515	1,883	—
403	401	1	391	80
19,133	14,706	4,427	17,327	128
2,443	1,212	1,231	2,238	364
6,272	243	6,028	5,967	—
2,023	488	1,535	1,634	274
1,052	527	526	3,206	2,466
2,248	58	2,190	1,765	—
1,391	235	1,157	1,258	38
623	169	454	1,531	984
9,933	716	9,216	8,631	—
3,335	64	3,270	2,956	—
4,572	345	4,227	4,236	—
1,921	307	614	1,362	—
105	—	105	87	—
7,037	1,787	5,251	5,186	408

Source: U.S. Bureau of the Census, *Survey of Governments 1979–1980* (Washington, D.C., 1981) p. 35.

spending on highways, health facilities, and higher education. The remaining categories are, for the most part, a
local responsibility, with local governments outspending
state governments by more than $3 to $2. This does not,
of course, indicate that states are not paying their share.
The state role has increased significantly during the past
decade and is likely to increase during the next decade as
federal aid shrinks. About 60 percent of state budget expenditures flows to localities or directly to individuals.
Local governments—counties, municipalities, and special
districts—will remain the primary "spending unit" for
public capital investments.

Investment by
Type of Project

During the last decade, state and local capital priorities have changed significantly. State and local capital
spending on sewers, transit, and water projects have increased more rapidly than other categories (table 4), largely as a result of increases in federal funding for selected
programs. The relatively slow growth of expenditures in
highways during the 1970s—a category that had grown
rapidly in the 1960s—was a result of the nearing comple-
tion of the system and the slow growth in trust fund revenues because the federal government had not increased
the four-cents-per-gallon gasoline tax since 1962. In addition, some highway construction funds have been diverted
to finance transit (table 5). Highway expenditure has also
shifted from new construction to a greater emphasis on
resurfacing, restoration, and bridge repair:

> During the latter half of the 1970s, federal capital support
> likewise assumed a geographic purpose. Many of the new
> programs—Federal Aid Urban (Road) Systems, Community
> Development Block Grants, Urban Development Action
> Grants, Mass Transit funding—were especially directed to
> older cities and their requirements for capital renovation
> (Peterson and Miller, 1981 p. 51).

These programs, however, are likely to be scaled back
and some may be turned back to the states entirely under
the president's proposal for a new federalism. Since pat-

Table 4

Capital Outlays by State and Local Governments by Type ($ in millions)

Fiscal Year	Local Schools	Highways and Bridges	Sewers	Transit	Water	Total
			Function			
1971	$4,845	$11,888	$1,744	$ 446	$1,247	$33,137
1972	4,759	12,317	2,091	435	1,343	34,237
1973	4,856	11,459	2,428	920	1,435	35,257
1974	5,108	12,152	2,640	926	1,743	38,084
1975	6,532	13,646	3,569	1,203	2,111	44,817
1976	6,547	14,209	3,955	1,339	2,208	46,531
1977	5,982	12,497	4,208	1,573	2,302	44,896
1978	5,709	12,898	4,366	1,407	2,141	44,769
1979	6,370	15,567	5,619	1,618	2,701	53,196
1980	7,362	19,133	6,272	1,921	3,335	62,894
Percentage Growth, 1971–1980	52%	61%	260%	331%	167%	90%

Source: U.S. Department of Commerce, Bureau of the Census, Governmental Finances, Fiscal Years 1971–72 through 1979–80.

terns of public capital spending seem to have reflected federal rather than local priorities, the planned withdrawal of Washington will require a radical rethinking of present practices and the reestablishment of local priorities.

The Fiscal Capacity of State and Local Governments

Capital spending by state and local governments must be viewed against the broader background of state and local revenues and expenditures. Revenues provide the resources to finance bond issues and maintain capital projects. In the past decade, state and local governments have sought to diversify their fiscal bases by introducing new taxes and charges, and relying less on traditional tax revenues. Trends in expenditures by category reveal changes in public service priorities which must be considered in making public investment decisions. In general, rapid increases in health and welfare programs have squeezed capital investments.

Recent Trends in Revenues

Overall state and local revenues nearly tripled between 1970 and 1980—with state government revenues growing faster than those of local governments (table 6). Growth was most rapid during the first half of the decade and actually declined in real terms since 1977 as the result of legislative and constitutional limitations on revenue sources, increases in federal aid, and draconian attempts to cut spending. There were, however, pronounced shifts in the relative importance of different revenue sources (table 6):

☐ The dominant revenue source of state governments continued to be the sales tax. However, it accounts for 8 percent less of the share of total revenues in 1980 when compared with 1970. Individual and corporate income taxes increased their share of state revenues from 18.8 percent to 23.7 percent.

☐ Property taxes remained the most important revenue source for local governments but experienced a 13 percent decline in their share of total revenues.

☐ Service charges of local governments experienced a 3 percent increase in share, while the share of charges in state revenues declined.

☐ The most dramatic change has been the tremendous increase in transfers to state and local governments from the federal government, from $21.8 billion in FY 1970 to $83 billion in FY 1980.

☐ State grants to local governments also grew rapidly, from $27 billion in FY 1970 to $81 billion in FY 1980.

From the perspective of financing public works, these shifts are important. Historically, the property tax provided revenues that backed local bond issues, either out of current revenues or out of the incremental revenues attributed to the project. The decline in importance of property taxes has not occurred voluntarily. In many cities and counties, property tax rates have reached constitutional or legislated ceilings, and so local jurisdictions have sought politically more expedient sources. Second, the rapid growth of federal revenues has been reversed and is projected to decline at an annual real rate of 14 percent during the next two years. Third, basic revenues to state governments have failed to grow as rapidly as expenditures. The ACIR summarized the problem: "Without the help of periodic increases in state sales and income taxes, most of our 50 state-local systems cannot generate sufficient automatic revenue growth to keep pace with, much less exceed the rate of growth of the economy" (ACIR, 1981, p. 8). The property tax revenues to local governments have increased more rapidly than the general rate of inflation, in part due to the inflation-induced tax benefits of property ownership. If inflation of property values abates, the reverse will be true—the property tax base will grow much more slowly than the local economy and local government need during the next few years.

Expenditures

State and local expenditures also grew rapidly during the 1970s, although, since 1975, they have declined in real terms—in large part as a result of restrictions on revenues (tables 7 and 8). Summarizing the trends depicted in these tables:

☐ The spending levels of state and local governments nearly tripled between 1970 and 1980. Reflecting their greater ability to raise revenues, state government expenditures grew at a slightly faster rate than localities. State governments rapidly increased their level of spending for public welfare, while decreasing their share of contributions for highway and roads. State matching requirements for federal public assistance and decreases in federal highway aid accounted for most of the state shifts. Education continued to be the largest single expenditure category.

☐ Local governments also continued to spend a major portion of their budgets on education. Outlays for sewers rose sharply in response to the federal funding of 75 percent of costs. Local spending for police protection also grew slightly. Local spending for public welfare, historically considered a local issue, declined

Table 5

Outlays for Federal Capital Assistance
($ in millions)

Fiscal Year:	1970	1971	1972	1973	1974	1975	1976	TQ[a]	1977	1978	1979	1980
Interstate Highways	3173	3330	3342	3269	2909	2804	3306	828	2828	2614	3163	3998
Interstate Resurfacing, Restoration & Rehabilitation	—	—	—	—	51	66	337	216	23	94	151	193
Interstate Transfers for Transit	—	—	—	11	35	170	342	109	392	667	700	679
Federal Aid Urban System	—	—	—	—	—	—	—	—	434	472	640	871
Bridge Replacement and Rehabilitation	—	—	3	19	38	46	104	35	131	146	208	548
UMTA Sec. 3 Discretionary Grants	1	285	510	864	870	1197	1092	254	1250	1400	1226	1655
UMTA Sec. 5 Formula Grants	—	—	—	—	—	152	390	55	611	735	1134	1120
Capital Uses of Sec. 5 Grants	—	—	—	—	—	9	25	7	39	N/A	251	431
Waste Water Treatment	—	—	—	680	1560	1940	2429	918	3545	3194	3741	4343
Community Development Block Grants	—	—	—	—	—	38	983	439	2089	2464	3161	3902
Local Public Works	—	—	—	—	—	—	—	—	575	3041	1720	416
Urban Development Action Grants	—	—	—	—	—	—	—	—	—	N/A	N/A	255

Note: 1. Outlays for FY 1965–70; $681 million. a refers to Third Quarter 1976 when federal fiscal year changed from July to October.

Source: Agency records from the Federal Highway Administration, the Urban Mass Transportation Administration, the Environmental Protection Agency, the Department of Housing and Urban Development, and the Office of Management and Budget; compiled in Peterson and Miller, 1981, p. 51.

Table 6

State and Local Government Revenues by Sources, 1970 and 1980 ($ in millions)

Item	State				Local			
	1970		1980		1970		1980	
	Amount	Percent	Amount	Percent	Amount	Percent	Amount	Percent
General Revenues	57,507	(83.7)	169,266	(79.6)	51,392	(86.2)	130,027	(83.4)
Taxes	47,962	(69.8)	137,075	(64.5)	38,833	(65.2)	86,387	(55.4)
Property	1,092	1.6	2,892	1.4	32,963	55.3	65,607	42.1
Individual Income	9,183	13.4	37,089	17.4	1,630	2.8	4,990	3.2
Corporation Income	3,738	5.4	13,321	6.3	—	—	—	—
Sales & Gross Receipts	27,254	39.7	67,855	31.8	3,068	5.1	12,072	7.7
Other Taxes	6,695	9.7	15,917	7.5	1,173	2.0	3,718	2.4
Current Charges	6,102	8.9	16,545	7.8	8,770	14.7	27,828	17.9
Miscellaneous	3,443	5.0	15,646	7.4	3,788	6.4	15,812	10.1
Utility, Liquor & Insurance	11,185	16.3	43,370	20.4	8,165	13.7	25,845	16.6
TOTAL Own Source Revenue	68,691	100.0	212,636	100.0	59,557	100.0	155,873	100.0
Intergovernmental Transfers	20,248		64,326		29,525		102,425	
From Federal	19,252		61,892		2,605		21,136	
From State	—		—		26,920		81,289	
From Local	995		2,434		—		—	

Notes: Figures in parentheses [( )] are not included in cumulative percent total. (Ed. refers to Intergovernmental transfers can be omitted)

Source: Buckley, Michael Patrick, *Assessing the Issues and Trends in Public Facilities Financing: Planning and Policy Considerations for State and Local Governments in Oregon,* University of Oregon, Masters Thesis, May 1982, various tables.

by nearly 30 percent when compared to its 1970 share of total expenditures.

☐ Both states and localities increased their spending share on debt interest, with the state level undergoing a 50 percent jump, from 3.1 percent in 1970 to 4.7 percent in 1980.

State and Local Fiscal Capacity and Public Investments

During the 1980s, the combination of a weak national economy, high interest rates, and the delegation of greater fiscal and administrative responsibility by the federal government has created serious problems for state and

---

Table 7

State and Local Government Expenditure, as a Percent of Gross National Product, Selected Years 1949–1981

Calendar Year	Total State and Local Expenditures[1]	Federal Aid	State-Local Expenditure From Own Funds	
			State	Local[2]
1949	7.8	0.9	3.4	3.5
1959	9.6	1.4	3.8	4.4
1969	12.6	2.2	5.3	5.1
1974	14.3	3.1	6.0	5.2
1975	14.9	3.5	6.2	5.2
1976	14.7	3.6	6.1	5.0
1977	14.0	3.5	5.7	4.8
1978	13.9	3.6	5.6	4.7
1979	13.5	3.3	5.6	4.6
1980	13.5	3.4	5.6	4.5
1981 est.	12.9	3.0	5.5	4.4

Notes: 1. Derived from National Income and Product Accounts. Numbers include federal aid.
2. The National Income and Product Accounts do not report state and local government data separately. The state-local expenditure totals (National Income Accounts) were allocated between levels of government on the basis of ratios computed from data reported by the U.S. Bureau of the Census in the annual governmental finance series.

Source: Advisory Commission on Intergovernmental Relations, *Significant Features of Fiscal Federalism, 1980–81 Edition,* Washington, D.C., 1981, p. 15.

Table 8
State and Local Government Direct General Expenditure by Function, 1970 and 1980 ($ in millions)

Function	State				Local			
	1970		1980		1970		1980	
	Amount	Percent	Amount	Percent	Amount	Percent	Amount	Percent
Education	13,780	28.2	35,251	24.5	38,938	47.2	97,960	43.8
Transportation	11,395	23.4	21,381	14.9	6,603	8.0	15,942	7.1
Public Welfare	8,203	16.8	33,242	23.1	6,477	7.8	12,310	5.5
Health & Hospitals	4,788	9.8	15,666	10.9	4,880	5.9	16,507	7.4
Police Protection	688	1.4	2,060	1.4	3,806	4.6	11,433	5.1
Sewerage	–	–	334	.2	2,167	2.6	9,558	4.3
Sanitation	–	–	–	–	1,246	1.5	3,322	1.4
Parks & Recreation	–	–	1,274	.9	1,888	2.3	5,247	2.3
Housing & Urban Renewal	23	.0	331	.2	2,115	2.6	5,731	2.6
Governmental Administration	2,060	4.2	6,840	4.8	3,908	4.7	11,595	5.2
Interest on Debt	1,499	3.1	6,763	4.7	2,875	3.5	7,984	3.6
Other	6,313	12.9	20,476	14.2	7,679	9.3	26,132	11.7
TOTAL Expenditure	48,749	100.0	143,718	100.0	82,582	100.0	223,621	100.0

Source: Buckley, Michael Patrick, Assessing the Issues and Trends in Public Facilities Financing: Planning and Policy Considerations for State and Local Governments in Oregon, University of Oregon, Masters Thesis, May 1982, various tables, p. 31–32.

local governments, problems that are unlikely to abate in the immediate future. The combined state and local sector faces a negative operating balance in 1982 (figure 1). They have neither carried-over surpluses nor the promise of federal assistance as they did in the recession of 1974–75. Fiscal resources are also undermined by the burgeoning underground economy—estimated to be as much as 15 or 20 percent of GNP (and growing)—which is placing more and more economic activity beyond the reach of state and local tax collectors.[5] In addition, some state and local governments have reduced their tax base by giving tax abatements and exemptions to attract industry without carefully weighing the costs and benefits. Yet deteriorating public capital may deter local industrial development more than tax incentives may encourage it.

The problems created by shrinking tax bases have been made worse by the proliferation of constitutional and legislative constraints on tax increases. Eighteen states now face constitutional or statutory limits—fourteen of these constraints have been enacted since 1978 (table 9). Twenty-nine of them limit states' local property tax rates, and nineteen limit property tax revenues—eleven enacted since 1978. The result is that states and localities have much less flexibility to respond to fiscal problems today then they did in 1975. A recent analysis by Steven Gold (January, 1982) concluded that limitations at the state level have not proved an active constraint on spending levels in large part because of the sharp decline in revenues caused by the recession. With more responsibility delegated from Washington, some states will find constitutional and legislative constraints binding as revenues recover.

These constraints on revenues are not balanced by any real constraints on the expenditure side of the increasingly harder-to-balance equation. School district spending is limited in five states, municipal spending in three, and county spending in two (table 9). Most observers feel, and many state and local government officials have argued, that the sharp recession in 1975 and the growing restlessness of taxpayers during the last five years have sweated much of the fat out of local budgets, reflected in the decline of state and local spending as a percent of GNP, from 15

FIGURE 1

State and Local Operating Balances, 1960–1982

$ Billions

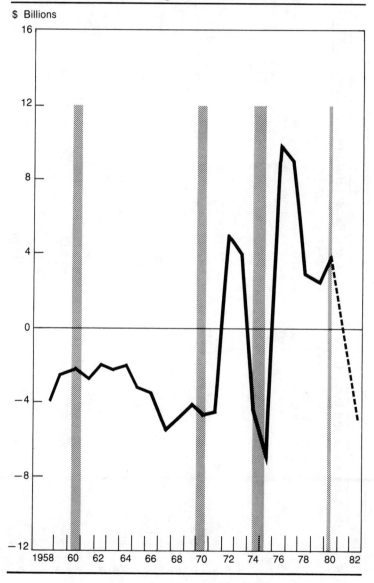

Table 9
Restrictions on State and Local Government Tax and Expenditure Powers (June 1, 1981)

States	Overall[1] Property Tax Rate Limit	Specific[1] Property Tax Rate Limit	Property Tax Levy Limit	General Revenue Limit	General Expenditure Limit	Limit on Assessment Increases	Full Disclosure	Limits on State Governments
(State Imposed Limits on Local Governments)								
Total	14	29	19	5	6	6	9	18
Alabama	CMS***	CMS*						
Alaska	CMS**							
Arizona	CMS***	CMS*	CM**					Const.***
Arkansas			CM***		CMS***	CMS***		Const.***
California	CMS***		CMS***[2]		CMS***	CMS***		Const.***
Colorado		CS*	CM*				CMS***	Stat.**
Connecticut								Const.***
Delaware		CS**	C***[2]					
Dist. of Col.								
Florida		CMS*					CMS**	
Georgia		S*						
Hawaii							C**	Const.*** Stat.***
Idaho	CMS***	CMS*	CMS***					
Illinois		CMS*						
Indiana		CMS*	CMS***					
Iowa		CM*	CM**		S**	CMS**		
Kansas					S*		CMS***	
Kentucky	CMS*							
Louisiana		CMS**	CMS***[2]					Stat.***
Maine								
Maryland	CMS***				CM**	CM**		
Massachusetts	CS*		CMS***					
Michigan		M*	CMS***					Const.**
Minnesota		CMS**		CM**				

State					
Mississippi		CMS*			
Missouri	CMS*	CMS***		CMS**	Const.***
Montana	CMS*				
Nebraska	CMS*	CMS*** / CM**			Stat.***
Nevada	CMS*				
New Hampshire					
New Jersey	CMS**	C**			Stat.**
New Mexico	CMS*	CMS***	MS**	CMS**	
New York	CM**				
North Carolina		CMS*** / CMS**			
North Dakota					
Ohio	CMS*				
Oklahoma	CMS*	CMS*			Stat.***
Oregon	CMS*	CMS*	CMS***		Stat.***
Pennsylvania	CMS*				Stat.**
Rhode Island					Stat.***
South Carolina	CMS*				
South Dakota		3			
Tennessee	CMS**			CMS***	Const.***
Texas	CMS*			CMS***	Const.***
Utah				CM**	Stat.***
Vermont					
Virginia		S**			
Washington	CMS**	CMS**	S**		
West Virginia	CMS*				Stat.***
Wisconsin	CMS*	CM**			
Wyoming	CMS*				

C—County   M—Municipal   S—School District   *—Enacted before 1970   **—1970 to 1977   ***—1978 and after
Const.—Constitutional   Stat.—Statutory

[1] Overall limits refer to limits on the aggregate tax rate of all local government. Specific rate limits refer to limits on individual types of local governments or limits on narrowly defined services (excluding debt).
[2] Limits follow reassessment   [3] Limit followed transition to a classified property tax.

Source: ACIR staff/C. Richardson based on data from CCH, Steven Gold, IAAO, and state survey.

33

percent in 1975 to below 13 percent in 1982. Many factors will continue to propel demands for state and local public services upward while fiscal resources will not grow proportionately. A recent report by Merrill Lynch concluded that:

> The planned Reagan Administration cutbacks in federal expenditures bear the greatest challenge to the credit standing of certain state/local governments and the market for their tax-exempt securities since the Great Depression (Herships and Karvelis, 1981, p. 1).

Future Demands for Public Infrastructure[6]

In spite of tight budgets, state and local governments must wrestle not only with how to maintain the current stock of public capital but with how to develop the new infrastructure that will be required by the transformation of the national economy. The most dramatic forces that will influence demands for public investments include:

☐ *Technological Changes.* Scientific advances promise to transform the structure of the national economy and the relationship between where we live and where we work. Coincident advances in data processing and communications technology are dramatically reducing the costs of storing, processing, and transmitting information. The development of new materials will lead to the replacement of structural steel by more durable compounds using silicon and carbon. Biogenetics promises significant breakthroughs in medicine and agriculture. Increased public investments in research facilities and communications systems will be required.

☐ *Demographic Changes.* The population and labor force will grow much more slowly in the 1980s than in the 1970s. By 1985, the labor force growth rate will be half what it was in 1975. The rate of growth of households will exhibit a similarly sharp decline. Housing investments will be less important, while education and training investments will have to be increased. The "graying of America" will propel soaring expenditures on health and social service facilities.

☐ *World Trade.* The growing importance of world trade will create rapid declines in employment in import substitute goods and services and a rapid expansion in goods and services for export. There will be severe local economic and fiscal dislocation as a result. Private investments to enhance competitiveness in world markets will have to be complemented by public investments in transportation and distribution systems.

☐ *Resource Management.* The 1970s were dominated by rapid increases in oil prices. Conservation and the development of alternative energy sources mean that sharp real increases in oil prices are unlikely during the next decade. However, the consequences of poor management of many resources, including water, topsoil, and the safe disposal of hazardous waste, will require large-scale public investments during the next decade.

Each of these changes will alter the structures of state economies, will shift the relationship between place of work and place of residence, and will change the demand for public infrastructure. States must develop a much-longer-term planning horizon in order to anticipate accurately what types of public investments must be undertaken and where they will be needed. Failure to understand the process of development and its implications for public policies can convert problems into crises. We are still recovering from the urban crisis, the energy crisis, and the reindustrialization crisis. Each could have been ameliorated by analyzing the issues and designing appropriate policies. Public investment policies must be guided by a careful look at the future.

Conclusions

During the next decade, state and local governments must make rapid and fundamental changes in their capital investment strategies. Traditional revenue sources are declining, while the need to increase capital spending is becoming more pressing. State and local governments must recognize and define the limits on public sector responsibility and establish new priorities—priorities that reflect not only the constraints on state and local fiscal capacity, but that also reflect the underlying structural shifts in our economy.

CHAPTER II NOTES

1. The issue of public and private responsibility is discussed in chapter 3, and, in more detail, in volume 1 of this series, *Planning and Managing.*

2. See Holland, 1972 and Abt, 1980. Also, see note 2 in chapter 1.

3. In August 1982, *Newsweek* featured a cover story on the deterioration of the nation's public capital; *U.S. News and World Report* did the same in September 1982; and the *National Journal*, in November 1982.

4. Recently, the U.S. Department of the Interior has begun disinvesting in natural resources, by turning land back to state governments and selling some land to private developers.

5. Estimates are hard to make, see Fiege, 1980.

6. These issues are discussed in more detail in volume 1 of this series, *Planning and Managing.*

CHAPTER III

Guidelines for Financing
Public Capital Investments

DECIDING HOW TO FINANCE public capital investments entails answering three questions. The first is,
Which responsibilities properly belong to the public sector? State and local governments must determine whether
they intend to continue full financial and administrative
responsibility for the services and subsidies they currently
provide. They must also rigorously question new demands
for public activities—convention centers, business development projects, or even new parks.

The second question is, What is the most efficient way
of meeting public sector responsibilities? Economists refer
to this choice as selecting the efficient mix of inputs—land,
labor, energy, and capital. The selection should be based
upon the relative price of the inputs. For example, as energy costs rose during the past decade, local governments
used extra labor and capital to make public buildings more
energy efficient. Now that capital costs have increased—
because of high interest rates—ways must be found to supply public services through capital-efficient techniques.
For example, instead of building a new laboratory for a
community college to teach electrical engineering, a cooperative arrangement could be made with a local manufacturing firm to allow training in the plant.

The third question is, How should required capital expenditures be financed? There are two issues: first, How
should costs be shared between the public sector and private developers? and second, What fiscal resources should

be used to pay for the construction and maintenance—debt, general revenues, or fees?

The discussion in the following chapters addresses the last of these three questions. (The first two are discussed in detail in volume 1 of this series, *Planning and Managing*.) This chapter outlines, briefly, how all three questions must be addressed in order to establish a strategic framework for public capital investment policy.

Public and Private Responsibilities

After a century of almost continuous expansion, the last few years have witnessed significant retrenchment of federal, state, and local government activity and employment. In the mid-nineteenth century, state and local governments were responsible for the construction of streets and sidewalks, the provision of police protection, and public education. Other responsibilities were added as a result of public outcry or disaster. Sanitation, fire protection, zoning, and utility regulation expanded municipal control over the social and economic environment. The Great Depression rapidly accelerated the federal government's involvement in direct income distribution and job creation. Since World War II, public sector activity has grown on four fronts: increased services for the poor, including social security, health care, housing, training, education, and day care; regulation of the economy, including environmental policy, worker safety, health, and capital markets; health, mortgage finance, and recreation; and economic development policy, including tax expenditures and private project finance. Economic development policies undertaken during the 1970s most obviously crossed the border that has traditionally divided public and private responsibilities. State and local governments now provide private, for-profit firms with low-interest loans, tax abatements, subsidized production facilities, wage and training subsidies, custom-built transportation links, low-cost water and water treatment, and subsidized energy.

Public services vary among states and localities reflecting historical accident, constitutional and legislative constraints, and local economic conditions. By and large, the gradual transfer of responsibility from private to public hands has been ad hoc with little consideration for the long-run consequences, and with no reference to any guidelines.

Prodded by increasingly tight budgets since 1975, state and local governments have begun the painful process of reconsidering what services and subsidies they can realistically afford to provide. Community centers and libraries are being closed. Developers are asked to build their own streets and sidewalks. Transit fares are raised. Program cuts are being made in the same ad hoc fashion that prevailed during expansion. As governments make these retrenchment decisions, the political strength of constituents is a more important consideration than whether the service is better conducted by the private or public sector. The Reagan administration has accelerated the process by cutting back or terminating many categorical grant programs and by converting others into block grants. "New federalism" would precipitate an even more radical surgery of programs and activities by state and local governments.

Changed circumstances call for a new strategy. States can no longer deal with periodic budget crises by across-the-board cuts in all programs. Further cuts in many programs will seriously jeopardize their usefulness. Nor can states expect, as in the past, to maintain programs by vigorous lobbying in Washington. Instead, state and local governments must determine which services can be turned over to the private sector or terminated. States must redefine their responsibilities if they are to be able to pay for those activities that need direct state government involvement.

Assigning priorities for public spending will require a rigorous questioning of traditional practices. If a convention center is a demonstrable boon to a local economy, why is it not financed privately?[1] Are public subsidies to business activity really necessary? Must a recreation site be developed and operated with public funds? Is public ownership of the water supply system necessary when electricity and gas are privately provided? Does an airport need to be owned by the county? Some economists have proposed radical measures such as the outright sale of subway systems and toll bridges to private firms, replacing public schools with education vouchers, privatizing garbage collection, and selling off state parks. The economics and politics of these proposals are complex and will require careful research and analysis by all levels of government. Little is known about the income distribution implications. Unfortunately, privatization has become a partisan political issue and therefore has not been discussed objectively as an integral part of a public investment strategy.

The basic guideline for state and local officials to follow in deciding "who does what" is that the public sector should only intervene if market-determined outcomes are neither efficient nor equitable. Even if public intervention is required, it does not have to involve spending public funds or public ownership of facilities. Regulation of private suppliers—of drinking water prices, for example— can lead to the desired supply of services without full public ownership. The task is by no means easy, but without cutting back on some public services, there will not be enough tax revenues to pay for those services that properly belong in the domain of state and local government.

Efficient Production of Public Services

Having decided what services to provide, policy makers will next have to determine the level of services and how to supply them. These decisions should be made along the same lines that a for-profit company determines how to invest—by comparison of costs with benefits. Unfortunately, benefits are much more difficult to measure for publicly provided services than they are for private investments. Sales revenues provide a private firm with a rapid check on the profitability of a product or service. In spite of the difficulties, however, it is possible to measure the value of public services. There is a growing literature that shows how to use the revealed preferences of consumers to measure the demand for public services. For example, changes in land property have been used to measure the benefits of highway noise abatement programs and visit rates can measure the implicit demand for a state park. (This issue is discussed more fully in *Planning and Managing.*) Comparison of projected benefits with projected costs should be an efficient gauge of whether a public investment should be undertaken. Yet, in planning public investments, many public agencies rely on engineering standards based upon the durability of materials, traffic flow criteria, design and use specifications, or, even inspired guesswork. While these guidelines may indicate where to anticipate structural problems, they do not directly indicate the costs and benefits of specific patterns of maintenance and investments.

If a public facility is fully financed through user fees (or dedicated tax revenues that are equivalent to user fees), planners will have a much clearer measure of the value of the project (chapter 4). Comparison of projected revenues from user fees with projected construction and operating costs indicates whether the project is worth undertaking. User fees have another benefit—they ensure that the service is used efficiently. A pricing system allocates a scarce resource to those who are willing and able to pay for it. Although user fees cannot be used for all public investments, they can be used much more broadly than at present.

Assuming that the appropriate *level* of a public service has been determined, the next decision is how to provide it most efficiently at the lowest possible cost. What is the appropriate input mix? This involves choosing among alternative "production techniques" based upon the relative costs of capital, land, labor, and energy. The pursuit of efficiency would dictate that relative input costs be the only factors considered by local planners, but political reality has often dictated otherwise. For example, capital expenditures on transit rolling stock could be reduced by hiring more maintenance staff and so avoiding the need to replace equipment as frequently. Yet, the Urban Mass Transit Administration was much more generous with capital grants than with operating subsidies. The result, according to several studies, was that many cities spent too little on maintaining and repairing buses and subway cars and too much on the purchase of new equipment. Because of the design of federal subsidies, the "lowest-cost" approach of local government agencies was not the "lowest-cost" solution for society as a whole. Similarly, waste water treatment construction grants subsidized only the construction of facilities, not their operation. Several studies have found that this has led to drastic over-building of plants, reflecting both the availability of free federal construction funds and the desire to minimize operating costs for which the local government is fully responsible.[2]

Within economists' definition of *efficiency*, deferred maintenance of infrastructure is not necessarily inefficient or even short-sighted. If the cost of capital is high relative to other factors, it may be cost-effective to allow roads to deteriorate below established engineering standards and to delay rehabilitating public office buildings. And if public demand for a facility or the service it provides has fallen, disinvestment may be appropriate. We need to develop a framework for analyzing public investment decisions that will lead to more efficient, and, therefore, lower-cost planning and budgeting to meet public service demands.[3]

Principles for Financing Capital Investment

Where grants from a higher level of government are not available, state and local governments must turn to three sources of funds to finance the construction and maintenance of public capital projects. First, they can borrow, and service the debt either from general revenues and the full faith and credit of the issuing jurisdiction (General Obligation bonds, GOBs) or from dedicated tax revenues or user fees (Limited Obligation bonds, LOBs). Second, they can finance the projects directly out of tax revenues or charges and incur no debt. And third, they can require or encourage the private sector to incur at least part of the costs directly.

There is no single efficient or equitable way of funding a public works project, and even if it were possible to identify the most efficient way to finance a project, the legal and institutional capabilities vary greatly among states and among local jursidictions. What may work in one county may be constitutionally prohibited or institutional-ly impossible in another. This does not preclude us from devising some general principles to guide an effective and equitable financing strategy.

The first principle is that, wherever possible, *those who benefit from a public facility—a road, park, hospital, or port—should pay for its development and operation, and the amount they pay should be related to the level of their use*. In short, those who reap the benefits should bear the costs. However, it is not always possible, through a simple "public pricing" mechanism, to ensure that all those who benefit from a particular project or facility actually pay for it. For example:

☐ Firms and households living downstream from a water treatment plant will enjoy cleaner water as a result of the facility.

☐ Retail stores may experience a boom in business after a nearby bus or rail service is opened.

☐ Commuters on congested highways will be able to get to work more quickly if mass transit services attract other commuters out of their automobiles.

These are examples of what economists refer to as *externalities—benefits enjoyed by those who are not direct users of the facility. In these cases, revenues from other sources should be used to supplement direct user fees.

There are other reasons for not relying completely on user fees. Direct fees may be very expensive to collect. For example, installing tolls on every street and road to pay for road maintenance and street lighting would be absurd. But even if direct user fees are not possible, taxes can be designed to approximate user fees (see chapter 4).

The second principle is that *the cost of a public capital project should be amortized over the life of the project and the maintenance and operating costs should not be deferred*. This principle naturally follows from the first, since it ensures that there is no intergenerational transfer of net benefits or net fiscal burden. If a long-lived project is financed out of current revenues, then future generations will enjoy the benefits with none of the costs. Conversely, financing a facility with an expected economic life of seven years with twenty-year bonds will shift the cost

43

to future generations and leave the jurisdiction with a future tax burden higher relative to its fiscal and economic base than at present. The failure to pay for adequate maintenance in past decades has transferred a fiscal burden of hundreds of billions of dollars from past generations onto the shoulders of today's taxpayers.

Third, *the operating and maintenance expenses associated with a project—operating a convention center, maintaining a bridge, or repairing and upgrading a resource recovery center, for examples—should be explicitly considered when designing the financing package.* Earmarked revenue sources must be sufficient to cover both debt service and ongoing expenses. Otherwise, unanticipated and often large subsidies from general revenues will be necessary. Too often, project plans contain swollen rhetoric concerning the jobs that will result and the general tax revenues that will be generated if the project is undertaken, but make no provision to ensure that the projected benefits are realized. For example, the operating subsidy for a convention center built from a state or local bond issue could be met from specific revenues identified in advance—such as a hotel room occupancy tax, or a part of the increase in sales tax and local real estate tax revenues. This protects the city or the state from entering into an open-ended, and often rapidly escalating, subsidy out of general revenues. This insulation can be strengthened if the ownership and operation of the facility are in the hands of a public authority—although excessive insulation can reduce public accountability (see *Planning and Managing*). This form of tax limitation can increase accountability and discourage continued public subsidy of a project that never becomes viable.

Fourth, *fiscal and administrative responsibility for a public investment project should be limited to those jurisdictions where most of the impacts are experienced.* For example, a program for maintaining roads and highways should be financed by a statewide tax—on gasoline or vehicle registration, for example. If county governments were individually responsible for their roads, then the transportation network would be a fragmented patchwork of roads that was not in the state's economic interest. On the other hand, a waste disposal facility that serves only one or two counties should be financed and run by those counties. In this instance, the state may serve a regulatory function to ensure compliance with uniform environmental standards. Adhering to this principle reduces the tendency to finance pork-barrel projects at the state level—voting for a state-financed convention center in one city in return for a state-financed industrial park in another.

This principle fails to recognize differences in need and fiscal capacity among jurisdictions. A locality may seek state financing of a facility on the grounds that it has an unemployment rate well above the state average and cannot aford to finance the project without state aid. Rather than helping poor areas by financing specific projects, states might consider less categorical assistance that allows localities to select which capital investments to make. A low-income jurisdiction might be helped more efficiently through revenue-sharing grants from the state rather than project-specific assistance. Most states operate some form of revenue sharing with their component jurisdictions, although few consider need and fiscal capacity when allocating the funds. This approach to local assistance allows localities to determine whether a project is of sufficiently high priority to compete with the other local demands and rely less on ad hoc grantsmanship in the state. If the project is 100 percent state funded, the locality does not have to make that painful decision. If the indirect approach of revenue sharing is not attractive, there are two alternative ways of recognizing local needs. States can engage in negotiating investment strategies with their local areas, or they can provide partial assistance in financing certain categories of investment—with the fraction paid for by state funds increasing according to measures of need and fiscal capacity.[4]

Ideally, all infrastructure financing decisions would be made in adherence to these four principles. In practice, however, decisions are made on political as well as economic grounds. Nevertheless, setting out the broad guidelines through which state aid will be provided is a necessary first step in the development of a long-term public investment strategy. Once the principles are publicly espoused, then the necessary exceptions will be clearly seen as exceptions, and the arguments to justify them will have to be more convincing. Local leaders and state legislators will be more prepared to accept a refusal from the governor to finance a pet project if they know that similar projects will not be financed in neighboring counties and if they can see that the refusal is a result of a consistent policy and does not reflect the governor's view of their political status or of the long-term commitment of the state to the development of the community.

CHAPTER III NOTES

1. Even if the convention center itself is not a viable undertaking, some of the private beneficiaries of convention activity, such as hotels and restaurants, could provide an annual subsidy to the company managing the center.

2. See U.S. Department of Commerce, 1980; American Public Works Association, 1981; U.S. GAO, 1982; Pagano and Moore, 1981.

3. A detailed discussion of cost-benefit evaluation and its use in policy analysis is contained in volume 1 of this series, *Planning and Managing.*

4. Chapter VI describes a framework for negotiating investments between state and local governments over major new projects.

CHAPTER IV

User Fees
for Public Projects

A PUBLIC WORKS PROJECT—from a statewide road system to a local library—is a commitment to pay for the construction, operation, and maintenance of the facility. This chapter outlines potential revenue sources to finance these costs in ways that meet the guidelines established in the previous chapter. The analysis emphasizes the imposition of user fees. The term *user fee* is used broadly to include any fee, charge, or dedicated tax that is paid by those benefiting from a facility or the services it provides. For example, user fees would include a gasoline tax where the revenues are dedicated to highway construction and maintenance, tuition charges at state universities, or admission fees to state parks.

How prices for public facilities and services encourage efficiency in both investment and in use is discussed in the first section of this chapter. The second section discusses some of the difficulties of imposing user fees, including how they affect low-income households, influence economic development, and may impede the local budget process. The final section outlines how user fees can be structured for different types of public investments including highways, health and education facilities, ports and terminals, recreation facilities, water supplies, water treatment plants, and waste disposal facilities.

User Fees and Efficiency

User fees should not be regarded as fiscal gimmicks to circumvent local revenue or expenditure limitations.

If properly designed, they can serve the same function that freely determined prices serve in private markets for private goods—providing incentives that promote efficiency in both the production and consumption of goods and services. User fees are, in the words of the late Selma Mushkin (1972), "public prices for public goods." Prices set at the correct level by the supplier of the goods or service (that is, at the marginal cost) serve a dual purpose. They encourage consumers to choose efficiently between alternative goods and services, and they signal to the supplier whether to expand, reduce, or maintain the existing level of output. Setting efficient public prices is therefore a valuable component in an effective public investment strategy.

Not all dedicated tax revenues are user fees. For example, Colorado had dedicated 50 percent of the net revenues from its new state lottery for capital projects. While this type of dedicated tax does guarantee a revenue stream for public works, it does not encourage efficiency in use and production that would result from a user fee.

Public policy makers frequently ignore the powerful influence that efficient pricing can have on consumption. When OPEC ended the era of low-cost oil in 1973, many experts claimed that rising prices would not reduce the quantity of oil demanded. A complex set of conservation subsidies and allocative regulations was introduced instead of allowing oil and gas prices to rise. Yet experience over the last nine years has taught us that consumption is very responsive to price changes.

Experience with user fees suggests that they are less likely to encounter as strong opposition as increases in general taxes. Since they are dedicated to specific activities, they are less likely to be diverted to other purposes during periods of fiscal stringency. In Cincinnati, for example, user fees cover 80 percent of the cost of the municipal water supply system. In spite of severe cuts in the city's operating budget during the 1970s, its water system was much better maintained than those in most comparable cities (Humphrey, Peterson, and Wilson, 1979). In general, infrastructure is better maintained in cities that support related services through dedicated user fees than in those that rely on general revenues (Pagano and Moore, 1981; U.S. Department of Commerce, 1980).

State and local governments collected nearly $40 billion through user fees in FY 1977 (table 10). By comparison, they raised about $70 billion through property taxes. Revenues from user charges could be expanded to finance public investments and to allow cuts in property taxes and other general revenue taxes.

The purpose of charging for the use of public facilities is that, if properly designed, user fees ensure efficiency in both production and use because prices send signals to both consumers and suppliers. For example, if consumers of water must pay a price equal to its marginal cost (to the municipality for supplying the water), they will weigh the value of consuming an additional gallon of water—for irrigating crops, washing mined minerals, or for watering lawns—against its price. They may choose to conserve water or consume it, but if they use water, they pay for what they use. Without water metering, artificial "plenty" is introduced, and local users have no incentive to economize on its use. An efficient user fee, therefore, encourages conservation without resorting to cumbersome, and often highly inequitable, tax subsidies of the type passed in many states to encourage energy conservation. Consumers are usually able to judge for themselves whether a particular conservation technique is cost-effective for their residence, factory, or farm. Areas that have installed water meters have experienced sharp declines in per capita water consumption. Consumers may take time to adjust to prices—many homeowners waited several years to determine that oil prices had permanently increased before installing insulation, for example. But we have mounting evidence of the effectiveness of prices in shaping consumer behavior.

User fees will also affect the "fairness" of the distribution of benefits and costs of public services. For example, when water is unmetered, intensive users (who are often high-income homeowners with large gardens) are subsidized by less-intensive users, usually those with lower incomes.

Most discussions of user fees have focused upon the

behavior of consumers—the users of public facilities. But prices should also guide agencies managing the facilities. For example, if, at an efficient price for water, the quantity consumed is reaching the capacity of the local system, it may be time to add new wells or reservoirs and treatment plants. Conversely, if revenues from a convention center fail to cover even operating costs, then the local government should seriously consider closing down or selling the facility.

A user fee does not ensure a more efficient use of a public facility unless it is set at a level that reflects the *marginal,* i.e. the incremental, cost of providing the service. Yet, most public officials aim at covering costs when imposing user fees—that is, they select a price equal to the *average* cost (total costs divided by the number of users). In many instances the average cost is not the efficient price. Only if the costs of production are fairly constant over a wide range of output are average and marginal costs equal, and the appropriate price will be the average cost. For example, the costs of accommodating visitors at a state park may rise proportionately with the number of visitors. But if average costs rise with increased output, then the "marginal cost" price will be above the average cost. For example, a waste disposal facility operating near full capacity may be able to handle additional material only by placing its staff on overtime or greatly increasing its per ton energy consumption. If it charges its marginal cost for all the waste it accepts, its total revenues will exceed total costs, which will provide a profit. There is nothing wrong with a public enterprise earning profits. Locally owned natural gas utilities earned an aggregate profit of about $60 million in FY 1976–77 (table 10). However, most publicly owned utilities tend to sell their output at below the market price, often by block discounts to large customers. The result is a subsidy to favored consumers.

Services that require very large capital expenditures and incur only low operating expenses, such as sewer systems or treatment plants, will exhibit marginal costs well below average costs. In these cases, efficiency calls for an operating subsidy paid out of general revenues.

Efficient pricing also requires consideration of the pattern of demand over time. If a public facility is used at a constant rate throughout the year—such as a waste disposal facility—then setting a price is relatively simple. However, most public facilities experience very high use during certain hours of the day or during certain months of the year. Bridges and subway systems are crowed during commuter rush hours and have excess capacity at other times. A park or municipal beach may be heavily used during summer months and deserted at other times. Museums may be full at weekends and much less frequented during the week. Efficient pricing for these facilities requires a two-tier, or even a multi-tier, pricing system. The economies of peak-load pricing are complex. During peak hours, days, or even months, the price should cover both operating and capital costs. During off-peak hours, the price should cover only operating costs. Thus transit fares would be higher in rush hour than during off-peak hours. Admission to campsites would be higher during summer weekends than at other times. Many private enterprises that experience systematic fluctuations in demand use a peak-load pricing system. Theater tickets are more expensive on weekends than during the week. Hotels that rely on the week-night business trade often offer discounts for weekend guests. Resort hotels charge a lower price out-of-season.

The economic rationale for the dual price system is that it is peak users that are responsible for the marginal capacity of the system. A transit system must purchase extra buses that are used solely during peak hours. Therefore these users should pay for the cost of that additional capacity. During off-peak hours, there is surplus capacity and users should pay only for the operation and maintenance of the facility or equipment.

The dual price system tends to spread the use of the service or facility more evenly and therefore reduces the overall capacity needs. For example, someone using the transit system for a shopping trip may be persuaded by the lower price to travel during nonpeak hours. A high summer rate on water use may discourage automobile washing during those months.

The benefits of peak-load pricing must be weighed against its costs. Peak-load pricing incurs additional administrative costs—metering and billing, for example—

Table 10

Role of Charges in Local Government Finance, 1976–77

($ in billions)

Category of Local Expenditure	Direct Expenditure	Charges
Education	75.7	3.5
Higher Ed. Institutions	4.8	1.0
School Lunch Sales (gross)	NA	1.6
Other	NA	.9
Highways	9.2	0
Welfare	11.9	0
Health	2.7	0
Hospitals	8.6	6.1
Police	8.8	0
Fire	4.3	0
Parks & Recreation	3.9	.6
Natural Resources	.9	.7
Sanitation & Sewage	8.8	3.1
Housing & Urban Renewal	3.2	.9
Air Transportation	1.1	1.1
Water Terminals Transport	.5	.4
Parking Facilities	.3	.3
Correction	1.6	0
Libraries	1.2	0
Financial Administration	2.2	0
General Control	4.5	0
Interest on Debt	6.3	0
Other & Unallocable	11.7	2.9
Water Supply	6.4	5.0
Electric Power	8.3	6.8
Gas Supply	.8	.9
Transit	5.0	1.7
TOTAL	192.7	37.5

Notes: NA means not available.
    a - through a gasoline sales tax.

| Charges As a % | Potential for Effective Strategy | | | |
Expenditures	None	Weak	Medium	Strong
5%		x		
21			x	
NA			x	
NA	x			
0				x[a]
0	x			
0		x		
71				x
0	x			
0			x	
16		x		
18		x		
35			x	
28			x	
100				x
80				x
100				x
0	x			
0		x		
0	x			
0	x			
0	x			
25	NA			
78				x
82				x
108				x
34		x		
19%				

Source: Bureau of the Census, *1977 Census of Governments,* Washington, D.C., various tables.

that may outweigh potential benefits. But there are many areas where it can lead to significant capital cost savings. It is regrettable that many municipal utilities price their services perversely. Water is offered at a discount per gallon for major users, even though they consume water during peak hours or weeks. Municipal electric power companies often provide power to commercial firms operating between 9:00 A.M. and 5:00 P.M.—peak hours—at rates below those paid by residential users, even though residential use is more evenly spread throughout the day.

In summary, an efficient user fee system is much more than a source of revenues. It is a way of ensuring that scarce resources are not wasted, that facilities are used efficiently, and that the need for investments in public infrastructure may be reduced. It will also provide an invaluable guide to effective planning and budgeting for public capital investment.

Limits to User Fees

The suggestion of applying user fees to public facilities and services is often strongly opposed. Some of the objections can be justified on economic grounds, others are based on misinformation or upon the politics of user fees. The five most frequent arguments are:

☐ Publicly provided facilities and services bring widespread social benefits for which a price cannot be charged. For example, police and fire protection services cannot easily be charged for when they are used.
☐ The services provided by the local public sector are redistributive—that is, they are provided for those with low incomes who could not otherwise enjoy them —and therefore should be paid for by the entire community.
☐ Public facilities and services are necessary to attract development and therefore should be subsidized from general tax revenues. The fiscal surplus from the ensuing development, it is argued, will pay for the services in the long run.
☐ Dedicating tax revenues for specific purposes reduces the flexibility of the budget-making process to deal

with rapidly changing public priorities and needs, and reduces the accountability of local officials.

☐ Dedicated revenues and separate administration of public projects prevent a coordinated approach to public infrastructure development.

There is one barrier to the adoption of user fees that is not based upon economic and political considerations and that states are powerless to influence. It results from a flaw in the federal tax code. Individuals who file itemized federal income tax forms (about one-fifth of all taxpayers—mostly from higher-income households) may deduct state and local taxes from their federally taxable income. This shifts some of the burden of high state and local taxes from high-income state residents to federal taxpayers at large. There is no provision on federal tax forms for itemizing user fees paid to state or local entities. A switch from financing the water supply system out of general property taxes to a quarterly water bill based on meter reading will reduce the homeowners' deductions, therefore raising the cost of water (net of federal taxes) even if the actual amount paid to the municipality remains unchanged. There are changes in the federal tax code that could aid the shift from general revenue financing of facilities to user fees—such as a "supplementary" deduction that would be proportional to the share of state and local revenues that is raised through user fees. It is an issue that should be addressed as part of a national public works investment strategy.

Let us review the five issues listed above.

External Benefits

It is not always easy to identify the beneficiaries of public services. Some public activities provide benefits which are of such widespread benefit and so intangible that user fees would be wholly inappropriate. Police protection is of value to all residents whether or not they are the victims of crime. Fire protection provides a sense of security that is of benefit whether or not one's residence ever catches fire. General revenues are appropriate for these general benefits.

Most publicly provided goods and services provide some external benefits which cannot easily be recovered directly through user charges. Commuters who use mass transit rather than automobiles lessen traffic congestion and reduce air pollution—benefits enjoyed, but not paid for, by automobile commuters. Graduates from the local school system or from local training programs will be less prone to go on welfare or commit crimes—benefits enjoyed by the whole community. Economists argue that those using facilities or services yielding external benefits should be subsidized to encourage increased use. In some instances, these external benefits can be captured—by using a portion of gasoline tax revenues to support mass transit, for example. The existence of external benefits does not always preclude user fees, but it does mean that the type of fees imposed must be based upon a broad study of the benefits of the facility or service.

Income Redistribution

The free provision of public services such as museums, libraries, and garbage collection, and the heavy subsidies to mass transit, are often justified on the ground that the poor could not afford these services if they were not offered below cost. This is a weak argument. Only 13 percent of the urban population is below the poverty level. Although low-income households represent a disproportionate share of mass transit riders (higher-income households tend to use private automobiles) and of public school students (higher-income households are more likely to send their children to private schools), the poor use less than their population share of water, produce less garbage, and are only infrequent visitors to museums and libraries. The subsidy is not only an expensive way to help those in need but also removes the incentives for efficient use.

In some cases, it would be cheaper and more efficient to charge a user fee for these services and facilities coupled with a cash grant to the poor so that they can afford to use them. For example, low income households could be given "museum stamps," or "state park stamps" out of the revenues generated from user fees. The poor may be better helped through specific and targeted income supplements rather than through the mispricing of public facilities. In this way, the efficiency aspects are retained, while the undesired inequity aspects are offset.

If those with low incomes are to receive preferential treatment, then either those with higher incomes may have to pay more than their share, or the subsidy to the poor will have to be paid out of general revenues. Some user fees can be structured so that they are progressive— an ad valorem vehicle registration fee is a good example. Others clearly cannot be—an income-related water rate would be an administrative nightmare. User fees are not necessarily more regressive—that is, they fall disproportionately heavily on the poor—than are general revenues (sales, property, and income taxes). A progressive personal income tax structure can lose a great deal of its progressivity after all the loopholes for home ownership, consumer debt interest, and other preferential activities are included. A sales tax that is levied on food and prescription drugs but not on personal services is regressive. The question of whether the property tax is progressive or regressive has yet to be settled. Increased reliance on user fees, especially if special provisions for low-income households are included, may actually improve the equity of the state and local revenue structure as well as increase its efficiency.

The argument against this change is that either the compensating programs will not be implemented, or, if they are, they will be vulnerable to budget cuts later. Hidden subsidies to the needy, it is argued, are preferable to overt assistance. State and local officials are seen as protectors of the poor against the wishes of taxpayers, and the main protection tactic is deception. Such "protection" may be successful on some occasions but cannot be successful all of the time. Failing to impose user fees leads to deteriorating facilities and services that benefit neither the poor nor those able to afford full-cost pricing.

User Fees
and Economic Development

Many state and local governments construct special freeway linkages, offer low-cost water, or abate taxes in order to attract firms. Whatever user fees the company might owe are partially abated to lower the local business costs in the hope of inducing development. Yet, there is little evidence that these business incentives are effective

(Kieschnick, 1981; Vaughan, 1979). They result in higher taxes or reduced services to other businesses and residents, which may offset any benefits offered to the few fortunate companies. Economic development may be more effectively encouraged by pursuing a prudent fiscal policy and by clearly stating the development policies and priorities. A reliable and equitable system of user fee revenues, guaranteeing the maintenance of highways and roads, a high quality education system, or an adequate water supply, may prove a more compelling inducement to business growth than a policy that mortgages future tax revenues in order to bribe footloose firms. If the relocation of a business is expected to generate increased tax revenues in the future—through increases in property or business tax revenues—then the tax increment should be set aside to finance construction and maintenance of the necessary infrastructure. Tax increment financing is, in principle, a type of user fee.

Flexibility and Accountability

One argument against user fees or dedicated tax revenues is that, by tying up a large share of revenues, the flexibility of the annual budget process is reduced, and the accountability of state and local officials is weakened. Dedicated funds can be cut back or diverted for other purposes—but only by legislative action (at the state level) or council approval at the local level. The decision is therefore made publicly, and after public discussion. From this viewpoint, some loss of flexibility may lead to the development of a planning horizon more compatible with the time horizon of the public investment. One of the major factors that has led to cuts in public spending on infrastructure maintenance and development has been the short-run time horizon of local elected officials and the escalating complexity of state and local budgets. Some balance must be sought between a rigid earmarking of revenues and excessive flexibility that leads to the neglect of long-term investments.

Coordination

Dedicated revenues and fees tend to be channeled through individual departments or even public authorities. Since each agency has less need to negotiate its annual budget, it may be less inclined to cooperate and coordinate with other agencies. This problem can be addressed through the capital budgeting process, through strengthened planning activities, through a stronger cabinet government, or through various ad hoc mechanisms (see *Planning and Managing*).

Some General Principles For a User Fee System

The preceding discussion suggests some general principles for states seeking tax revenues to finance the capital and operating costs of public projects.

1. Taxes and user fees should be imposed on the major beneficiaries of the project or service.
2. These public prices should be set equal to the marginal, not the average, cost of providing the project or service.
3. Where there are systematic seasonal or hourly fluctuations in use, peak-load prices should be imposed during periods of heavy use and lower prices at other times.
4. Where the shift from general revenue financing to user fees will prove particularly burdensome to low-income households, special provisions should be used to ensure adequate access. This may require higher-income users to pay more than the efficient, marginal-cost price.
5. Revenue sources should be responsive both to inflation and the economic growth rate.

Having discussed some of the general issues associated with user fees, we now turn to an examination of how user fees can be used to finance the capital, operating, and maintenance costs of different types of public investments.

Opportunities for User Fees

The purpose of a system of user fees is to recapture sufficient benefits that accrue to users and other beneficiaries of public projects and services to cover both capital and operating costs in a way that promotes efficiency in use and production. In many instances, a single tax or fee may be insufficient, and a set of taxes will be necessary. For example, a convention center may bring additional business to local hotels, retail stores, local printing firms, catering companies, and equipment-leasing firms. It may also increase land values in the surrounding areas. Equitable recapture of these different benefits requires taxes on different types of business activity, on property values, and on any external benefits the center may create. Projects that serve broad public purposes may require some subsidy from general revenues.

Earmarking funds to pay for infrastructure capital and maintenance costs can be done within several different administrative frameworks. On the one hand, full fiscal and administrative responsibility for the project can rest with a state or local government agency. On the other, responsibility can be delegated to a special authority. Theoretically, the questions of how to finance and how to administer are separable. In practice, they are not. Many states face constitutional limitations on the taxes they can impose, the extent of earmarking allowed, and the powers that can be granted to local governments or to special authorities. These states must modify their strategies to comply with these constraints.

In this section, eight major categories of capital investment are discussed: road transportation, higher education, transit, water supply, waste water treatment, waste disposal, ports and terminals, and recreation facilities. The section concludes with a brief discussion of tax increment financing and how it can be applied to different types of projects. These do not exhaust the types of public investments made by state and local governments, but the discussions indicate how pricing mechanisms can be applied to public facilities.

Road Transportation

Building and maintaining highways, bridges, streets, street lighting, and traffic systems currently absorb 10 percent of state and local budgets. About one-quarter of the $36 billion in annual expenditures is provided through grants from the federal government. About $16 billion is raised through gasoline and motor vehicle tax revenues (although these revenues are not necessarily dedicated to road transportation programs). Some additional revenues are raised through tolls and miscellaneous charges. In this area, there are many opportunities to increase user fee revenues that can be used to repair and extend the road system. Most of these user fees are best levied statewide in order to minimize collection costs and to reduce the possibility of interjurisdictional tax cut wars. Depending upon how responsibility for road construction and maintenance is shared between state, county, and city agencies, the state can distribute some of the revenues to localities.

Vehicle Registration Fees. Most states charge a fixed annual registration fee for motor vehicles, usually graduated according to the size of the vehicle. These fees must be periodically raised to reflect inflation. An alternative is to make the annual fee proportional to the value of the vehicle. This automatically provides increased revenues during inflationary periods and is more progressive than a flat fee—higher-income households own more valuable automobiles than do low-income households.

Gasoline Tax. The federal government and most states impose a tax denominated in cents per gallon. With the hindsight gained from more than a decade of inflation and a quadrupling of gasoline prices, and ad valorem gasoline tax (which would be equivalent to a sales tax) would have yielded much greater revenues. The American Highway Association magazine reported in 1981 that forty states had increased or were contemplating an increase in gasoline taxes. Unfortunately those that converted from a flat tax to an ad valorem tax recently suffered both from the slow rate of growth of gasoline sales and the decline in gas prices. These declines are unlikely to persist, and, in the long run, conversion should produce a strong revenue flow. We have argued that a strong case can be made for using some of the revenues to subsidize mass transit.

Other Sales Taxes. The Federal Interstate Highway Trust Fund collects taxes on the sale of batteries, tires, and lubricating oil as well as gasoline. Most states impose general sales taxes on these items and could impose an additional tax, dedicated for road repair. It would be administratively simple. Other possible revenue sources could include a vehicle transfer tax and increases in fines for traffic violations.

Private Construction. Some highway spurs and even roads and highways are constructed for the benefit of a single private developer—a new factory or a large resource development project. Too often, state governments have been prepared to underwrite the full cost of the construction as part of an incentive package to lure the corporation. Where a single, primary beneficiary can be identified, that beneficiary should finance at least part of the costs (chapter 6 discusses these types of arrangements in more detail).

Bridge and Highway Tolls. Before the advent of the Federal Interstate Highway Trust Fund, many intercity highways and major road bridges in the Northeast and Midwest were developed using tolls to back bond issues and to pay for maintenance, usually administered through a public authority. The Federal Highway Program, which paid 90 percent of the construction costs, explicitly prohibits tolls on highways. Existing tolls on designated interstate highways that have accepted federal funds are scheduled to be phased out when bonds are retired. Yet, tolls are an efficient, if unpopular, user fee and have the advantage that they can be varied by time of day to discourage rush hour travel.

Consideration should be given to expanding highway tolls—including tolls on federally financed portions of the highway—and to developing peak-hour pricing on toll systems. Other innovations worth considering include commuter licenses displayed in automobile windshields allowing the licensee to use express lanes on highways or allowing access to specific highway segments during rush hour; highway-use meters installed in motor vehicles which would allow authorities to record time and use of highways of individual vehicles; and a sales tax surcharge on parking lots in designated areas during prescribed hours. Appropriate pricing for commuter access to the road system would encourage increased use of transit systems, reduce the rate of highway deterioration, and provide the revenues necessary to improve the quality and safety of road transportation.

There are other aspects of road transportation management that should be fully self-financed—from motor vehicle inspection and drivers licenses to highway patrol and traffic courts. The budgets of these administering agencies need not be supplemented by any general revenues, and charges could be adjusted automatically to cover operating expenses.

Institutions of Research and Higher Education

Universities, colleges, community colleges, and medical research facilities are a major state capital expense, and, increasingly, a central factor in shaping economic development. The rapid rate of technological change will require major public investments in new facilities and equipment if graduates are to learn the skills desired by industry. The growing demand for engineers, technicians, and physical scientists will be met only if the capacity of higher education institutions is expanded.

Nationwide, approximately one-third of the $40 billion spent annually by state and local governments on higher education are met through user fees, mostly tuition fees, which have been sharply raised in almost all states. Yet increased capital investments will only be possible if more revenues are raised to cover operating expenses. Tuition increases obviously impose special problems for the poor, especially in view of substantial reductions in federal grants and loan programs for higher education. Increases in state tuition costs should be accompanied by the introduction or expansion of state programs to assist low-income students.

In addition to tuition assistance, there are other innovative sources of revenue that can be used to finance public investments, higher education, and research facilities.

Currently less than 2 percent of university research is funded by private industry, while about 50 percent is funded by federal agencies. The remainder is financed by private foundations, state agencies, university endowments, and operating funds. Increased involvement by private corporations can help finance the construction of new facilities and the acquisition of modern equipment, as well as spur increased research efforts in areas that have the potential for commercial and industrial development.

State Matching Grants For R and D. The state of California has instituted the MICRO program that provides a proportional grant to universities that receive a grant from private firms to undertake research in the micro-processing field. The state of Pennsylvania has begun the Ben Franklin partnership which also matches private sector contributions. These programs do not necessarily require an increase in state spending, but, rather, they redesign the criteria through which funds are allocated and share the costs with private firms.

Patent Policy. In most states, patent royalties are simply regarded as general revenues to the state. Yet royalty revenues can be increased and earmarked for the expansion of R and D facilities. A more aggressive patent policy would have three elements. First, a substantial share of royalties should go to the inventor; second, the institution should be allowed to keep some of the revenues, perhaps to purchase research equipment and to encourage a more entrepreneurial approach to research funding. Finally, the remaining revenues could be earmarked to fund basic research.

Tax Policy. Some states provide tax incentives for corporate donations of equipment to educational institutions—following the federal tax incentive provided in the National Economic Recovery Tax Act. Universities should also be made aware of the advantages of leasing or contracting the services of equipment and facilities from private, for-profit corporations. These arrangements pass on to the institution the tax benefits of rapid depreciation and tax credits (see chapter 7). These measures can reduce the cost of equipment for research and higher education.

Cooperative Arrangements. The cost of R and D infrastructure can be reduced by cooperative arrangements with private corporations in which the firm offers research grants, equipment, and even the use of its own facilities in return for long-term research in specified areas conducted by the university's faculty and graduate students. State administrative procedures and university charters can discourage these arrangements. While there are legitimate concerns about academic freedom and about the rivalry between teaching and corporate research, a more aggressive state role can be negotiated that harnesses private resources to help finance public facilities.

Transit

The average, publicly owned, transit system covers less than 30 percent of its capital and operating expenses from farebox revenues. About half the gap is filled with federal capital and operating grants, and the remainder is met from state and local tax revenues. Supporters of transit subsidies argue that public transportation (1) is a necessary service for the poor—an argument that does not require such extensive subsidies, as we have seen; (2) is an important step in reducing our dependence on foreign oil—but high gasoline prices have proved a far more effective incentive; and (3) helps reduce traffic congestion. Some subsidies from federal, state, and local general revenues are appropriate but no empirical studies have concluded that they should be as large as present levels.

Mass transit systems do allow for a denser land use in urban areas and therefore increase the value of commercial and even residential land downtown. Therefore some case can be made for some subsidy financed from local property taxes. Overall, there is a strong argument for increasing transit fares—a move that has been made, amidst great publicity, in Chicago, New York, Boston, and other large cities. In no instance were special provisions made to assist poor households (although some systems do offer discounts to senior citizens).

To some extent, the problem of excessive subsidies for transit has arisen as a result of federal policies. When UMTA operating subsidies were first offered, only public systems that ran deficits were eligible. A local subsidy from general revenues could leverage increased federal funds while a hike in fares would reduce federal aid. Federal capital grants encouraged over-buying of buses and rolling stock and the construction of projects such as the Buffalo, New York, system that will not be able to cover more than a fraction of its operating expenses whatever fare is charged. However, cuts in federal aid are inevitable, and cities will now have to live with the expensive systems they lobbied so hard to win.

Unpopular though it may be, the efficient long-run solution to financing urban transit needs will require four steps. The first, which we have already discussed, is to increase fares. The second, no less difficult, is to cut back on public services. Some deficit-ridden routes and services might have to be abandoned. The third step is to encourage the growth of what have been called para-transit systems. These are new, low-cost, ways of providing transportation services. They may include jitney services—a hybrid between a bus and a taxicab—or community-owned, not-for-profit buses or dial-a-ride systems. Para-transit services will, in some areas, cushion the blow of the withdrawal of public transit services. Jitney routes would tend to develop in low-income neighborhoods where gypsy or pirate cabs now operate, often illegally. The fourth step is to remove or reduce the subsidies to automobile commuting such as toll-free bridges and under-priced downtown parking. If drivers had to pay the true cost of getting to work, more would become riders, and the escalating needs to expand highway capacity would abate. Innovation in service delivery and pricing can reduce the level of public capital investments needed.

Water Supply Systems

In *America in Ruins,* Pat Choate and Susan Walter estimate that refurbishing municipal water supply systems will require between $75 billion and $110 billion over the next twenty years, a level of expenditure that is more than four times the current level. Yet, less than half the water used by residential, commercial, and industrial customers is metered. Even where water is metered, the rates charged often do not reflect the marginal cost of supplying the water. Nonmetered users pay either through a flat assessment to a local special district or through general property and sales taxes to the local municipality.

In addition, fees and charges for hooking up a new user to the distribution system are often below the cost of the actual construction involved. The result is often waste in water use and implicit subsidies to new sprawling development. The massive subsidies for utility hook-ups are not restricted to water systems alone but apply to electricity, gas, and sewage systems. Until it raised charges in the late 1970s, the state of Vermont charged less than $20 for a new link to the electricity network, although the actual cost averaged more than $1,300. After the passage of Proposition 13 in California, the city of Petaluma imposed an impact fee of $3,000 per acre to developers in order to finance utility hook-ups. Other jurisdictions have imposed development fees as high as $4,000 per bedroom to cover utility system costs for new residential subdivisions (Peterson and Miller, 1981). There is no reason why the cost of water and other utility linkages should not be fully paid by those needing them. It has mistakenly been argued that development fees of this sort will raise the cost of new residential and commercial property and will therefore slow growth. But this is not true. The cost of housing is determined by demand and supply in a very broad metropolitan market in which any new construction is a tiny fraction of increased supply. Housing prices will, therefore, remain unchanged while the value of unimproved land will decline. Fees would therefore depress the value of land for residential uses, which may reduce land speculation at the urban fringe.

Areas not metering water should seriously consider installing meters. Where there is substantial excess supply capacity and very little seasonal fluctuation in demand (in urban areas in the Northeast), the costs may outweigh the benefits. In other areas, where water is scarce, metering may help reduce the need to expand capacity. The issue is often complex. In some area, municipalities, farmers, and mining interests compete for the same artesian water, and ownership rights are not clearly defined. In the long run, however, meeting growing demands for water will be possible only through a system of prices. State governments will, necessarily, play a larger role over time, assisting local governments in developing fee systems, financing water projects, and regulating private suppliers. For many states, these are unfamiliar functions, but functions that will be needed as part of an overall capital investment strategy.

Waste Water Treatment

Capital and operating costs for waste water treatment plants will grow rapidly during the next decade as local governments seek to comply with federally mandated water standards. In the past five years, the federal government has provided over $20 billion in grants to state and local governments for the construction of facilities, although these grants are now being scaled back (chapter 2). Localities are required to operate these plants through the imposition of user fees, although relatively little attention has been given to the structure of these charges.

While household waste water cannot easily be metered, the volume and concentration of industrial, commercial, and agricultural waste water can be measured. Many European countries already use fairly sophisticated waste water charges—with the rates set according to both the volume and type of pollutants. The topic is complex and will require extensive research and experimentation. State governments will be called upon to work with local governments and sanitary districts in developing consistent and realistic fees to finance waste water treatment.

Hazardous Waste Disposal

The consequences of unplanned and unregulated disposal of hazardous wastes are only just being understood. Love Canal has become a national symbol for the problem. Congress has appropriated money to a superfund to finance the clean-up of a growing list of waste dumps that threaten public health and safety. There is still no viable solution to the disposal of either high- or low-level nuclear waste.

Some of the problems have been caused by illegal dumping. Others have arisen from accidental spills. But many have arisen because, when the waste was dumped, the full environmental and health implications were not known. There are probably many more chemical and biological time bombs as yet undiscovered in landfills, lakes, rivers, and oceans.

It is the potential for as-yet-undiscovered hazard that makes it impossible to develop a private-sector solution to the waste disposal problem. No private firm wishing to set up a waste disposal site could obtain adequate insurance to cover potential health damages to employees or nearby residents. Anyone who would be harmed would have to establish negligence on the part of the company, which may be difficult and costly. These unusual risks will make it necessary for state and local governments to provide the facility and to guarantee residents compensation for any potential damages, regardless of whether the facility was operated negligently. The facility may be run by a private company under contract, but site selection, preparation, and regulation will have to be a state government responsibility.

For many industries—especially those developing new technologies—a modern disposal site for hazardous waste, operated under the auspices of the state government, will be a strong location incentive. Waste disposal will be an economic development inducement.

The cost of developing these sites need not fall on local taxpayers. Waste disposal facilities can be totally self-financing, with users paying for each ton or gallon of waste they bring to the site—a disposal fee usually less costly to a firm than the possibility of protracted lawsuits. The fee should not be viewed as a license to pollute. It could vary by type of waste and reflect the full cost of treating or storing the substance. It would, in fact, encourage companies to minimize waste production through using less-toxic materials, recycling, or treating the waste at their own plant.

Ports and Terminals

Marine ports, airports, and even trucking terminals are often publicly owned—usually administered through special public authorities. Most of their costs are covered through direct user fees. Some large ports, however, do not fully cover their operating costs and are subsidized from general revenues from city and state governments. Based upon general trends in transportation, these subsidies would seem geared more to preserving the historic past and favoring established interests than to investing in the future of an industry. The subsidies have been provided to protect or generate jobs, but the evidence of their effectiveness is no more than that for tax incentives or subsidies of industrial parks. Some profitable transportation facilities could be turned over to private ownership, particularly if substantial new investment or rehabilitiation is required, to take advantage of the tax benefits (see chapter 7). There are some opportunities to increase user fee revenues. For example, general aviation is subsidized by commercial airlines and by the federal and local governments. An aviation fuel tax could be imposed to finance that fraction of airport costs attributable to serving private planes.

Recreation Facilities

There are numerous potential revenue sources to finance recreation facilities that have been poorly exploited. From docking fees at municipal marinas to charges at state-run campsites, revenues rarely pay for the full cost of the facility. Public golf courses charge much less than comparable private courses and must rely on queuing to allocate access to the course. Fees for concessions are often left unadjusted for years, when an annual auction would be more efficient and yield more revenues. Yet these are all types of user fees that can be expanded.

Admission fees are not always feasible for parks, but there are alternative user fees. The development and maintenance costs of an urban park could be financed, in part, by a special assessment on nearby residents. Tax increment financing is particularly suitable as a source of revenues for new urban recreation projects—from parks to pedestrian malls—because these projects enhance property values (commercial and residential) within a limited and identifiable location. The technique does not have to be limited to property taxes. A successful downtown plaza will draw additional trade to retail establishments, generating additional sales tax revenues that can be earmarked to maintain the area.

Many studies have documented the serious under-investment in recreational open space, yet, with so many pressing and mandated demands on the municipal budget, recreation programs are invariably cut. Increased revenues paid by beneficiaries would ensure more adequate funding.

Tax Increment Financing

One mechanism for capturing the benefits of a public project is the earmarking of increases in general revenue taxes—tax increment financing. A new park will increase the property values of adjacent residences. A new convention center will increase hotel occupancy rates and the volume of business for service firms. A new or improved transit service will increase the values of nearby commercial and residential property. These increases in sales or property values lead to increases in sales, business income, and for the local and state government, property tax revenues—revenues that could pay for the maintenance of the facility or to service the debt incurred in its construction.

But not all tax revenue increments represent a fiscal bonus to jurisdictions. A large part will reflect increased demands for public services. For example, a new state park or transit route that leads to new residential development on land that was previously agricultural will certainly increase tax revenues but will also require additional public facilities and services. Very little of the tax increment should be used to pay for the original capital investment. At the other extreme, a public project in an already developed area—the conversion of abandoned inner city industrial space into a park or commercial center—may have very little effect on the demand for public services, and therefore most of the tax increment can be dedicated to meeting project costs.

There are many instances of tax increment financing—almost all have used increments in property tax revenues. In California, after Proposition 13, many towns and counties have used the technique to pay for infrastructure for new residential development. However, most of these financings have set aside the full increment in tax revenues to back bond issues and have made no provision for increased expenditures on public expenditures. In Buffalo, an improvement district has been proposed to pay for the maintenance of a commercial mall developed in conjunction with the new subway system. The district will impose a special tax on adjacent commercial properties.

Conclusions

This chapter has only touched upon the possibilities of funding the construction and maintenance of public facilities through the imposition of user fees. Many regulatory agencies can be financed through license application fees, charges for rate hearings, and other similar policies. Grazing rights on state lands can be determined by competitive bidding and the revenues used to pay for the preservation of wilderness areas.

User fees could provide an additional $30 billion for state and local governments—about a quarter of this would come from increased gasoline taxes and motor vehicle registration fees, a quarter from utility hook-up fees and water rates, and the rest from other sources. A part of these revenues could be used to finance the maintenance and development of public facilities. Part could be used to provide the resources to make up for reduced federal aid. The remainder could be used to reduce other taxes, or, more realistically, to reduce increases in other taxes.

The state's role will extend beyond that of managing its own fiscal problems. It must work with local governments to tackle the complex technical issues in setting and administering public pricing systems. It may also have to use both carrot and stick to ensure that localities do not try to compete for industry through cutting public prices and to ensure the consistency of local actions with broad state goals.

User fees cannot be applied to all public services and facilities. Many services should be paid for out of general revenues. But many public activities can be more efficiently paid for by the beneficiaries. User fees encourage efficient capital budgeting and efficient facility use. Dedicating revenues does reduce budgetary freedom. But this freedom, exercised within the short time-horizon of local elected officials, has led to the neglect of public capital and to fiscally imprudent financing decisions. The environment necessary for sustained residential, social, and economic development is not created by investment decisions that are reversed, altered, or abandoned every few years.

Decisions will be more carefully considered if they involve the long-run commitment of public revenues for a clearly specified purpose.

CHAPTER V

Improving the Operation of the Bond Market

ABOUT ONE-THIRD of state and local capital investments are financed through issuing tax-exempt bonds.[1] In recent years, the cost of debt financing has risen, both in absolute terms and relative to federal and corporate debt. Some jurisdictions have been driven out of the bond market altogether. While many of the factors that have escalated the cost and the difficulty of public debt finance are beyond the control of state governments, there are state actions that could reduce the costs of issuing bonds and ensure a more orderly market. This chapter examines these actions including finance and budgeting assistance, state bond-banks, state guarantees for local issues, and limits on the use of tax-exempt debt.

The first section of this chapter provides a brief overview of the use of debt by state and local governments. The second section describes the market for tax-exempt securities and why the market has deteriorated. The final section describes steps that states can take to improve the operation of the bond market. It must be emphasized that these are incremental changes. And, while they may assist small, distressed communities, they will do little to improve the overall operation of the bond market. In the face of high rates and a narrow market, state and local governments will have to be much more rigorous in determining what types of investment merit tax-exempt financing.

State and Local Debt

During the last decade, state and local indebtedness has grown but not as rapidly as GNP. State debt has grown more rapidly than local debt, although state debt is less than one-fifth of total exempt debt issued. The interest costs of state debt have tripled as a share of GNP since 1964, while they have remained a constant share of GNP for localities. There are wide variations among states in the level of indebtedness, largely as a result of constitutional limitations. Among new issues, revenue bonds are now nearly triple general obligation bonds, and most of the growth has come from issues by statutory authorities rather than state or local governments. High interest rates and lender uncertainty about future interest rates have forced state and local governments into short-term financing.

Overall Levels of Debt

In 1981, total state and local debt outstanding reached $370 billion (table 11), about 13.3 percent of GNP, and about 10 percent of all medium- and long-term securities outstanding. Revenues from bond sales financed about 30 percent of state and local capital expenditures in 1980, a decline from 50 percent in 1970. Most of the decline in bond financing was a result of the sharp increase in federal grants, which, by 1980, had become the single largest funding source (table 12). State and local debt as a share of GNP has actually declined since 1964, as has gross federal debt—in part a reflection of reduced investment in public infrastructure, but also reflecting how rapid inflation has reduced the value of long-term indebtedness. State participation in the bond market has grown relative to municipalities. During the past fifty years, state debt has grown from only 16 percent of the level of local debt in 1929 to about 60 percent of that level today. By comparison, state debt grew by 180 percent during the 1970s, federal debt by 144 percent, consumer debt by 176 percent, and residential mortgage debt by 202 percent. The volume of new debt issued by state and local governments has increased much more rapidly than private sector debt and equity. In 1970, new corporate debt and equity issued

Table 11
Federal, State, and Local Debt, Selected Years 1929–1981
($ in billions)

Fiscal Year	Gross Federal Debt	Total State Debt	Total Local Debt	Gross Federal Debt	Total State Debt	Total Local Debt
		Amount			As a Percent of GNP	
1929	$ 16.9	$ 2.3	$ 14.2	16.9	2.3	14.2
1939	40.4	3.5	16.6	46.1	4.0	18.9
1949	252.8	4.0	16.9	96.6	1.5	6.5
1954	270.8	9.6	29.3	74.5	2.6	8.1
1959	284.7	16.9	47.2	60.4	3.6	10.0
1964	316.8	25.0	67.2	51.4	4.1	10.9
1969	367.1[1]	39.6	94.0	40.6	4.4	10.4
1970	382.6	42.0	101.6	39.8	4.4	10.6
1971	409.5	47.8	111.0	40.2	4.7	10.9
1972	437.3	54.5	120.7	39.3	4.9	10.9
1973	468.4[2]	59.4	129.1	37.8	4.8	10.4
1974	486.2	65.3	141.3	35.8	4.8	10.4
1975	544.1	72.1	149.1	37.5	5.0	10.3
1976	631.9	84.4	155.7	38.9	5.2	9.6
1977	709.1	90.2	167.3	38.0	5.1	9.4
1978	780.4	102.6	177.9	37.4	5.1	8.8
1979	833.8	111.7	192.4	35.4	4.9	8.4
1980	914.3	122.0	213.6	35.6	4.8	8.5
1981 Est.[4]	995.1	135.5	235.0	34.9	4.8	8.5
	Percent Distribution			Annual Percent Change[3]		
1929	50.6	6.9	42.5	—	—	—

1939	66.8	5.8	27.4	9.1	4.3	1.6
1949	92.4	1.5	6.2	20.1	1.3	0.2
1954	87.4	3.1	9.5	1.4	19.1	11.6
1959	81.6	4.8	13.5	1.0	11.9	10.0
1964	77.5	6.1	16.4	2.2	8.1	7.3
1969	73.3	7.9	18.8	3.0	9.6	6.9
1970	72.7	8.0	19.3	4.2	6.1	8.1
1971	72.1	8.4	19.5	7.0	13.8	9.3
1972	71.4	8.9	19.7	6.8	14.0	8.7
1973	71.3	9.0	19.7	7.1	9.0	7.0
1974	70.2	9.4	20.4	3.8	9.9	9.5
1975	71.1	9.4	19.5	11.9	10.4	5.5
1976	72.5	9.7	17.9	16.1	17.1	4.4
1977	73.4	9.3	17.3	12.3	6.9	7.5
1978	73.6	9.7	16.8	10.1	13.7	6.3
1979	73.3	9.8	16.9	6.8	8.9	8.2
1980	73.1	9.8	17.1	9.7	9.2	11.0
1981 Est.[4]	73.0	9.8	17.2	8.8	9.4	10.0

Notes: 1. During 1969, three government-sponsored enterprises became completely privately owned, and their debt was removed from the totals for the federal government. At the dates of their conversion, gross federal debt was reduced by $10.7 billion.

2. A procedural change in the recording of trust fund holdings of Treasury debt at the end of the month increased gross federal debt by about $4.5 billion.

3. The percent changes indicated for years prior to 1970 are annual average changes since the previous year shown.

4. Estimated.

Source: ACIR staff compilation based on U.S. Bureau of the Census, *Governmental Finances,* various years; Office of Management and Budget, *Special Analysis, Budget of the United States Government. 1982;* United States Treasury Department, *Treasury Bulletin,* July, 1981; and ACIR staff estimates.

totalled $38.5 billion, while new municipal bond issues
were only $18.1 billion. In 1979, $43.3 billion of new state
and local debt was issued, while new corporate issues were
only $42.2 billion.

Although public debt as a share of GNP has declined,
annual interest payments as a share of GNP have grown
since 1964 (table 13). State governments have been especially harmed by rising interest costs, primarily because
a much higher share of their debt has been issued more
recently and is in shorter-term maturities. In 1981, the

Table 12
Sources of Financing for State and Local Capital Expenditures

Item	1960	1965	1970	1973	1977
Federal Aid	20.0%	22.0%	22.0%	25.2%	43.3%
Long-Term Bonds	37.1	35.0	51.0	45.0	32.1
Other Local Resources	42.9	43.0	37.0	29.8	24.5
Ratio of Annual Tax-exempt Bond Issues to Annual Capital Outlays	50.2%	51.8%	59.5%	67.5%	118.7%

Source: George E. Peterson, "Capital Spending and Capital Obsolescence: Outlook for Cities," in The Fiscal Outlook for Cities, edited by Roy Bahl (Syracuse: Syracuse University Press, 1978), p. 53.

Table 13
Interest Payments on Government Debt, Selected Years 1929–1981
($ in billions)

Fiscal Year	Federal[1]	State[2]	Local[2]	Federal	State	Local
		Amount			As a Percent of GNP	
1929	$ 678	$ 95	$ 550	0.7	0.1	0.5
1939	941	129	534	1.1	0.1	0.6
1949	5,339	97	330	2.0	*	0.1
1954	6,382	193	525	1.8	0.1	0.1
1959	7,593	453	963	1.6	0.1	0.2
1964	10,666	765	1,590	1.7	0.1	0.3
1969	16,588	1,275	2,457	1.8	0.1	0.3

Fiscal Year	Federal[1]	State[2]	Local[2]	Federal	State	Local
		Amount		As a Percent of GNP		
1970	19,304	1,499	2,875	2.0	0.2	0.3
1971	20,959	1,761	3,328	2.1	0.2	0.3
1972	21,849	2,135	3,894	2.0	0.2	0.4
1973	24,167	2,434	4,351	2.0	0.2	0.4
1974	29,319	2,863	4,803	2.2	0.2	0.4
1975	32,665	3,272	5,511	2.2	0.2	0.4
1976	37,063	4,140	6,129	2.3	0.3	0.4
1977	41,900	5,136	6,257	2.3	0.3	0.3
1978	48,695	5,268	6,714	2.4	0.3	0.3
1979	59,837	5,790	7,197	2.5	0.3	0.3
1980	74,860	6,763	7,984	2.9	0.3	0.3
1981 Est.	90,600	7,700	8,800	3.2	0.3	0.3
		% Distribution		As a Percent of General Revenue[3]		
1929	51.2	7.2	41.6	19.2	4.8	10.9
1939	58.7	8.0	33.3	20.9	3.6	10.7
1949	92.6	1.7	5.7	13.2	1.6	3.7
1954	90.0	2.7	7.4	9.1	1.6	3.9
1959	84.3	5.0	10.7	10.0	2.5	4.6
1964	81.9	5.9	12.2	10.4	2.7	5.3
1969	81.6	6.3	12.1	10.2	2.6	5.4
1970	81.5	6.3	12.1	11.8	2.6	5.6
1971	80.5	6.8	12.8	13.2	2.9	5.8
1972	78.4	7.7	14.0	12.5	3.0	5.9
1973	78.1	7.9	14.1	12.4	3.0	6.2
1974	79.3	7.7	13.0	13.5	3.2	6.3
1975	78.8	7.9	13.3	14.7	3.4	6.5
1976	78.3	8.7	12.9	15.6	3.9	6.6
1977	78.6	9.6	11.7	14.8	4.2	6.1
1978	80.3	8.7	11.1	15.3	3.9	6.1
1979	82.2	8.0	9.9	16.1	3.8	6.1
1980	83.5	7.5	8.9	17.9	4.0	6.1
1981 Est.	84.6	7.2	8.2	18.8	**	**

Notes: *Less than .05 percent. **Data not available.
   1. Interest on the public debt. Data for 1929–1949 are administrative budget figures; for 1964–1981, unified budget figures.
   2. Interest on general debt.
   3. General revenue from own sources (before intergovernmental transfers).

Source: ACIR staff compilation based on U.S. Bureau of the Census, *Governmental Finances,* various years; Office of Management and Budget, *Special Analysis, Budget of the United States Government. 1982;* U.S. Treasury Department, *Treasury Bulletin,* various issues; and ACIR staff estimates.

level of state debt was about 60 percent of the level of local debt, yet annual interest paid by state governments was about 85 percent of the interest paid by local governments. Interest costs on debt represented 4 percent of total state spending in 1981 compared with just over 1 percent in 1964.

However, these aggregate data mask considerable interstate variations (table 14). Eleven states have combined state and local debt in excess of $2,000 per capita.[2] Nine states have per-capita indebtedness below $1,000.[3] Care should be taken in interpreting these data. High indebtedness is not necessarily a sign of fiscal imprudence or impending fiscal collapse any more than low indebtedness is a good sign. Generally, a high level of indebtedness reflects past investments in public capital—although the relationship is not straightforward because there are great differences among states in the amount of public invest-

Table 14
State and Local Debt Levels in the 50 States: FY 1980

	Debt Per Capita	Debt as Percentage of Own Source Revenues	Debt Interest as % of State Expenses	Debt Per $1,000 of Per Capita Income
Alabama	1,049.14	93.9	3.3	151
Alaska	10,098.03	77.6	8.6	900
Arizona	1,748.42	74.9	2.9	208
Arkansas	882.79	102.2	2.4	127
California	1,022.83	151.1	2.1	102
Colorado	1,315.76	107.0	2.7	144
Connecticut	1,862.07	70.3	5.9	184
Delaware	2,725.79	54.8	5.9	292
Florida	1,075.34	100.0	3.4	126
Georgia	1,135.58	98.3	2.8	149
Hawaii	2,260.25	72.1	6.0	245
Idaho	712.33	151.0	2.2	94
Illinois	1,338.49	100.2	4.1	137
Indiana	714.90	144.3	3.0	83
Iowa	787.82	166.9	1.7	90
Kansas	1,386.64	91.6	3.1	150
Kentucky	2,027.27	49.6	6.6	274

	Debt Per Capita	Debt as Percentage of Own Source Revenues	Debt Interest as % of State Expenses	Debt Per $1,000 of Per Capita Income
Louisiana	1,686.22	75.6	4.7	222
Maine	1,190.43	89.2	4.0	169
Maryland	1,779.07	83.4	4.2	191
Massachusetts	1,784.41	82.8	4.8	201
Michigan	1,200.87	122.2	3.3	127
Minnesota	2,129.13	74.3	4.1	240
Mississippi	845.24	112.1	2.9	137
Missouri	860.17	116.6	2.6	104
Montana	1,083.49	125.1	3.2	141
Nebraska	2,932.67	46.8	2.1	338
Nevada	1,645.41	87.2	3.8	156
New Hampshire	1,405.80	70.2	5.7	168
New Jersey	1,713.39	82.5	4.8	176
New Mexico	1,455.80	103.6	3.9	193
New York	2,648.31	69.6	6.7	291
North Carolina	751.94	130.7	2.4	102
North Dakota	1,264.62	117.5	3.0	154
Ohio	970.77	113.3	3.3	111
Oklahoma	1,115.42	105.3	2.8	131
Oregon	2,668.45	54.7	6.3	299
Pennsylvania	1,718.43	71.7	6.1	201
Rhode Island	2,142.49	62.4	6.6	252
South Carolina	1,073.32	91.2	2.4	152
South Dakota	1,642.95	69.8	3.6	220
Tennessee	1,315.59	71.2	3.9	179
Texas	1,452.07	79.9	3.6	165
Utah	968.12	120.4	2.7	135
Vermont	1,543.84	77.3	5.0	211
Virginia	1,096.77	103.7	3.7	128
Washington	2,814.70	50.8	3.1	294
West Virginia	1,665.18	64.5	5.1	226
Wisconsin	1,205.79	118.6	3.1	142
Wyoming	2,476.76	85.9	5.2	250
District of Columbia	4,165.98	41.4	5.1	394
U.S. Average	1,481.66	89.2	4.0	16.9

Source: Calculated from Bureau of the Census, *Survey of Governmental Finances, 1981,* Washington, D.C., 1981, various tables.

ment that was federally financed. In the case of New York, some of the debt was incurred financing operating expenses before the city's fiscal crisis led to a return to Generally Accepted Accounting Principles (GAAP). In other states, infrastructure has been paid for out of general revenues. What really matters, from the perspective of the long-term economic and fiscal health of a state, is whether the accumulated investments were made wisely, whether they have been adequately maintained, and whether voters, legislators, and officials are prepared to take any action necessary to protect the integrity of their past investments.

Type of Public Debt

Long-term debt can be divided into two types.

☐ *Full faith and credit*—sometimes known as general obligation bonds (GOBs). This debt is unconditionally backed by the credit of the issuing government, implying its ability to tax. It includes debt whose interest and amortization is financed from specific tax and nontax sources, but which represents a liability payable from any other available resources if the pledged sources are insufficient. Most GOB issues require voter approval.

☐ *Nonguaranteed debt*—sometimes known as limited obligation bonds (LOBs). This consists of debt payable solely from earnings of revenue producing activities, from special assessments, or from specific nonproperty taxes. LOB issues do not usually require voter approval. Industrial Development Bonds (IDBs) are an example of LOBs.

Nationally, nearly 55 percent of state debt was nonguaranteed in 1980, 37 percent was in the form of general obligation bonds, 6.2 percent was full faith and credit payable from specified nontax revenues, and 2.1 percent was short-term (Gold, 1981).

There are substantial variations among states. In only thirteen states—mostly those with high per-capita debt levels—did general obligation bonds account for more than half of total debt in 1981.[4] Nine states have no general obligation debt because of constitutional constraints.[5] Table 15 describes the constitutional limitations on state borrowing.

At the county and municipal level, growth of general obligation debt has been more constrained. Table 16 describes the constitutional and statutory limitations on local borrowing. Limited obligation debt has therefore been a more important part of debt expansion during the past decade. Partly because of defaults on municipal GOBs in the 1873–79 depression and during the 1930s, 80 percent of the states have local debt limits on GOBs, usually expressed as a percentage of assessed property values, and about the same number require a community to hold a referendum before issuing GOBs (Shaul, 1980). About 65 percent of states and localities have reached these self-imposed debt ceilings. LOBs (excluding IDBs) have grown from 16 percent of new issues in 1952 to nearly 40 percent today. Among all new tax-exempt issues, revenue bonds are now nearly three times the volume of general obligation bonds, while they were only half the volume in 1970 (Public Securities Association, 1981). Because of the ease with which LOBs can be issued and the very large number of political entities that can issue LOBs, their rapid growth contributed to the growing disorder and high-risk premiums in the tax-exempt market.

The Term Structure of Public Debt

Between 1975 and 1980, most state and local borrowing was long-term (over twenty-four months). Since 1980, the trend has reversed. In 1981, the volume of new long-term debt declined while the volume of short-term increased by 36 percent. This was a result of high interest rates and uncertainty concerning future interest rates. Lenders are unwilling to purchase long-term fixed rate assets—in the late 1970s, many had been caught by the sudden upsurge of inflation holding assets that yielded a negative rate of interest. Borrowers have been reluctant to commit themselves to long-term obligations in the hope that rates will have fallen when they have to roll over the debt. The increasing share of short-term debt makes budgeting less certain. (See figure 2.)

Table 15

State Constitutional Limitations on State Borrowing, 1976

State	No Limitations	Legislative Borrowing Power Limits			Exceptions to Limits			
		For Casual Deficits or Extraordinary Expenses Only	For Any Other Purpose	Referendum Required To Create Debt	Referendum Required To Exceed Limit	For Refunding	Limit May Be Exceeded: For Defense of State or Nation	For Other Purpose
Alabama		$3,000,000	(1)	(1)		x	x	
Alaska			(3)	x		x	x	
Arizona		350,000				x	x	
Arkansas				x				
California			$300,000		x		x	x
Colorado		100,000	50,000		x		x	
Connecticut	x		(7)					
Delaware	x					x	x	
Florida			(9)	x		x		
Georgia		(10)	(10)	(10)			x	
Hawaii			(11)			x		
Idaho			2,000,000	x	x			
Illinois	x	(13)		x			x	
Indiana		(15)					x	
Iowa		250,000			x		x	
Kansas			1,000,000		x			
Kentucky		500,000			x		x	
Louisiana	x					x		
Maine		(16)	2,000,000		x			
Maryland	x						x	x
Massachusetts	x						x	x
Michigan		(19)		x			x	x

State	Amount / Notes				
Minnesota					✕
Mississippi					
Missouri	$1,000,000	✕			✕
Montana					✕
Nebraska	100,000 (21)				
Nevada			✕		✕
New Hampshire					✕
New Jersey	(22)	✕	✕		✕
New Mexico	200,000 (21)	✕	✕		✕ ✕
New York					✕
North Carolina	(15) (23)	✕	✕	✕	✕
North Dakota	2,000,000				✕
Ohio	750,000		✕		✕
Oklahoma	500,000 50,000	✕	✕		✕
Oregon		✕	✕ ✕ ✕		✕
Pennsylvania	50,000	✕			✕
Rhode Island		✕			✕
South Carolina	(15)	✕			
South Dakota	100,000 21, 1		✕		✕
Tennessee					✕
Texas	200,000	✕	✕		✕
Utah					✕
Vermont	(21)				
Virginia	(28) (28)	✕ ✕			✕ ✕
Washington	(15) (29, 12)	✕			
West Virginia	(15)		✕		
Wisconsin	(21)			✕	✕
Wyoming	(21)	✕			✕

Source: Advisory Commission for Intergovernmental Relations, Significant Features of Fiscal Federalism, 1976–77 (Washington, D.C., 1977), p. 94–95.

Table 16

State Constitutional and Statutory Referendum Requirements for Local Government Issuance of General Obligation Long-Term Debt, 1976

State	Citation[1]	Referendum Required	Approval[2]	Remarks
Alabama	C	X	M	
Alaska	C	X	M	
Arizona	C	X	M	Only for debt in excess of the 4 percent limit.
Arkansas	C	X	M	
California	C-S	X	M	
Colorado	C-S	X	M	
Connecticut	.....	None required	.....	
Delaware	S	X	M	
Florida	C-S	X	M	
Georgia	C	X	M	
Hawaii	.....	None required	.....	
Idaho	C-S	X	2/3	
Illinois	S	X	M	
Indiana	.....	None required	.....	
Iowa	S	X	2/3	
Kansas	S	X	M	
Kentucky	C-S	X	2/3	
Louisiana	C-S	X	M	
Maine	S	X	M	Applies to municipalities only.
Maryland	C-S	X	M	Constitutional requirement applies to municipalities, statutory requirement applies to charter counties.
Massachusetts	.....	None required	.....	Except for debt issued by regional school districts in which case a referendum may be called by the towns comprising the district; in this event, simple majority approval is required.

State				
Michigan	S	X	M	Not applicable to school districts. Numerous statutory exemptions as to when applicable.
Minnesota	S	X	M	Does not apply to Minneapolis, St. Paul, and Duluth.
Mississippi	S	X	3/5	Only on petition of 20 percent of the electors for county bonds; 10 percent or 1,500, whichever is less, for municipal bonds.
Missouri	C	X	2/3	
Montana	S	X	M	If turnout is less than 40 percent of the electorate (30 percent for schools), the bond issue fails. If turnout for school bond issue is between 30 and 40 percent, 60 percent majority is required.
Nebraska	C-S	X	M	Fifty-five percent for school districts.
Nevada	S	X	M	
New Hampshire	S	X	2/3	Not applicable to cities or counties.
New Jersey	S	None required	. . . . . .	Except for debt issued by certain classes of school districts (simple majority).
New Mexico	C	X	M	
New York	S	None required	. . . . . .	Except for debt issued by certain classes of school districts (simple majority). Permissive referendum for most town and village issues.
North Carolina	C	X	M	Referendum is not required if the amount of issue does not exceed 2/3 of the net debt reduction for the preceding fiscal year.
North Dakota	C-S	X	2/3	Simple majority for county bonds; 60 percent for municipalities and school districts with over 5,000 population.
Ohio	S	X	M	
Oklahoma	S	X	3/5	Except that in the case of county hospital bonds a referendum is required on petition only (20 percent of the electors).

(Table 16 continued on next page.)

87

Table 16 (continued)

State	Citation[1]	Referendum Required	Approval[2]	Remarks
Oregon	S	X	M	Applies only to debt in excess of statutory limit up to specified maximum.
Pennsylvania	S	X	M	
Rhode Island	S	X	M	Optional.
South Carolina	C	X	M	Applies only to debt issued by cities and towns.
Tennessee	......	None required	....	Except that a 3/4 majority vote is required for issuance of general obligation industrial development bonds.
Texas	S	X	M	
Utah	S	X	M	
Vermont	S	X	M	
Virginia	S	X	M	Applies to county debt only. No referendum required in counties that elect to be treated as cities.
Washington	C	None required	....	Except for township debt (2/3 majority) and debt issued in excess of constitutional limits (3/5 majority).
West Virginia	C-S	X	3/5	Applies only to school districts and townships. No referendum required for county or municipal bond issues.
Wisconsin	S	X	M	
Wyoming	C-S	X	M	

Notes: 1. The citation is either the state's constitution (C), statutes (S), or both (C-S).
2. A simple majority (a favorable majority of 50 percent plus 1 of all votes subject to counting on the question) is indicated by "M"; where more than a simple favorable majority is required, the required percentage is entered.

This table deals only with referendum requirements that apply generally to general obligation debt issued by cities, counties, and school districts in each state. As in the case of debt limits, there are numerous exceptions and special provisions, particularly regarding debt issued by special districts and for specific purposes. No attempt has been made to treat those special provisions in this tabulation.

Source: ACIR staff with the help of state attorneys general or other state officials.

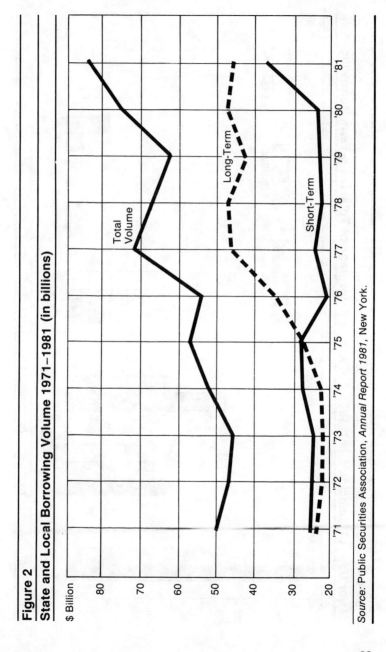

Figure 2

State and Local Borrowing Volume 1971–1981 (in billions)

$ Billion

Total Volume

Long-Term

Short-Term

Source: Public Securities Association, *Annual Report 1981*, New York.

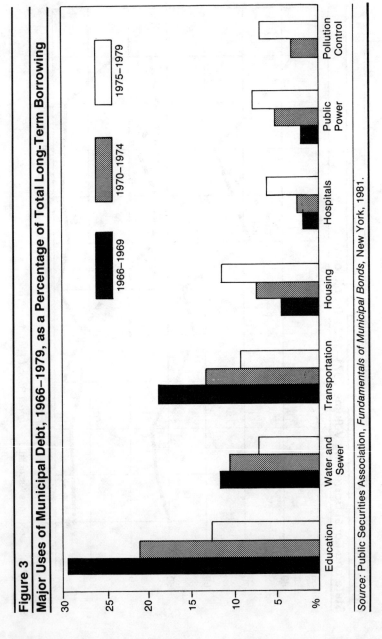

Figure 3

Major Uses of Municipal Debt, 1966–1979, as a Percentage of Total Long-Term Borrowing

1975–1979
1970–1974
1966–1969

Education
Water and Sewer
Transportation
Housing
Hospitals
Public Power
Pollution Control

Source: Public Securities Association, *Fundamentals of Municipal Bonds,* New York, 1981.

Table 17

Volume of Tax-Exempt Borrowing by Purpose
($ in billions)

Purpose of Issue	1970	1971	1972	1973	1974	1975	1976	1977	1978	1979	1980	1981[1]
Education	5.0	5.3	2.9	2.3	3.7	4.7	5.3	5.1	5.0	5.1	4.6	4.5
Transportation	3.2	4.3	1.7	1.6	1.5	2.2	3.4	2.9	3.5	2.4	2.6	3.5
Water & Sewer	2.4	3.2	1.3	1.9	2.0	2.3	3.0	3.3	3.2	3.1	2.9	2.9
Public Power	1.1	1.3	1.2	1.4	0.9	2.7	2.7	4.8	5.8	5.4	5.1	6.3
Pollution Control	NA	0.9	0.4	1.5	1.5	2.2	1.9	2.6	2.7	2.1	2.3	4.4
Hospital	NA	0.4	0.6	0.7	0.8	2.0	2.3	3.3	2.3	3.4	3.6	5.4
Indus. Revenue Bonds	NA	NA	NA	NA	NA	1.3	1.5	2.3	3.5	7.1	8.4	10.1
Housing	0.2	1.9	0.9	1.1	1.0	0.6	3.1	3.7	5.8	9.7	16.6	5.9
Total Long-Term**	18.1	24.9	23.7	23.8	23.6	30.7	35.4	46.7	48.1	48.7	58.3	47.7*
Short-Term Volume	17.9	26.3	25.2	24.7	29.4	29.0	21.9	24.8	21.4	21.7	27.7	—
TOTAL VOLUME	36.0	51.2	48.9	48.5	53.0	59.7	57.3	71.5	69.5	70.4	86.0	—

Notes: NA indicates figures not available
    1 Indicates totals for first 6 months
    * PSA Estimate
   ** Totals include Recreation, miscellaneous, and Refunding not broken out

Source: Francis Viscount, *Municipal Bonds: The Need to Regulate* (Washington, D.C.: National League of Cities), p. 9, from data in the *Daily Bond Buyer,* and in Public Securities Association annual reports.

Purpose of Debt Issues

The investment categories financed by debt have changed over the last decade (figure 3 and table 17). Of traditional public infrastructure purposes, education, transportation, and water and sewers halved their share of new issue revenues between 1966–69 and 1975–79, while housing, hospitals, public power, and pollution control investments (by private firms) doubled their share. Most of the increase in LOBs has been for investments that are not traditionally a public responsibility—mortgage subsidies, industrial revenue bonds, and pollution control. In 1980, Congress severely limited mortgage revenue bonds that had absorbed more than one-quarter of all the revenues from tax-exempt bond sales in 1980 (table 18). Those restrictions have been eased in the 1982 tax act and issues are again increasing. Industrial revenue bonds grew from only $1 billion in 1975 to over $10 billion in 1981 (table 17), prompting federal legislation to limit them somewhat in 1982, after a fierce and protracted battle with powerful lobbying groups.

Issuers of Tax-Exempt Debt

States play a small direct role in the issuing of tax-exempt debt. Only 10 percent of new issues in 1979 were by state governments, less than half their share a decade earlier (table 18). The share of counties in the total has remained fairly constant. The sharp decline in municipalities' share reflects the fiscal constraints and constitutional limits.

But states have been responsible for setting up the statutory authorities—from public utilities to port authorities—that now issue more than half of all new long-term tax-exempt debt. Unless the investments of these authorities are carefully coordinated, they may crowd out more traditional state and local capital investments.

The Tax-Exempt Bond Market

Interest rates on tax-exempt bonds have risen alarmingly since 1979, both in absolute terms and relative to

Table 18

**Municipal Borrowing by Type of Issuer,
Selected Years, 1966–1981**

	1966	1970	1974	1978	1980–81
States	21.0%	22.0%	15.9%	10.0%	11.0%
Counties	6.6	9.1	8.7	10.3	10.3
Municipalities	20.9	26.3	27.0	20.3	18.0
School Districts	14.2	11.8	9.2	5.4	4.0
Special Districts	6.9	6.4	5.3	2.3	3.1
Statutory Authorities	30.4	24.4	33.9	51.7	53.6

Source: Public Securities Association, *Fundamentals of Municipal Bonds* (New York, 1981), p. 51.

taxable bonds. Many issuing agencies have suffered lower credit ratings, and others have been scared away from issuing new debt. The problems have arisen from the increased volume of tax-exempt paper for sale, from reduced demand for tax-exempt securities because of changes in federal policies, and from the growing confusion in the operation of the market itself.

Tax Exemption of State and Local Debt

With one minor exception, all interest payments on bonds issued by state and local governments are exempt from federal income taxes and often from state and city income taxes as well.[6] This characteristic has established a somewhat separate capital market for state and local debt, set apart from federal and corporate long-term debt. Because investors are willing to accept lower yields to obtain tax-free income, this exemption lowers the borrowing costs of state and local governments—a subsidy paid by the U.S. Treasury. Before World War II, when personal and corporate income tax rates were low, this exemption led to a relatively small differential between taxable and tax-exempt bond rates. The interest on municipal bonds averages about 95 percent of the rate on taxable bonds. By the 1970s, however, dramatic increases in income tax rates—which increased the advantages of investing in tax-exempt bonds—had widened the differential from 5 percent to about 30 percent, a spread that had, in part, cushioned state and local governments from the full effect of rising interest rates (Peterson, 1982). In fact, if inflation is taken into account, state and local governments were able to borrow at *negative real* interest rates for most of the second half of the 1970s.[7] However, because of the increased volume of bonds and reduced demand, that trend has been reversed and the differential has narrowed from 35 percent in 1979 to below 20 percent in the first half of 1982 (Peterson, 1982).

Although tax exemption for public bonds does reduce the cost of public borrowing below the private sector rate, it only does so on a "before-tax" basis.

> Because private investors can deduct interest payments, the private before-tax borrowing rate is reduced by a percentage equal to the applicable marginal tax rate of the private borrower. Most private capital formation is carried out by the corporate sector, and since the corporate tax rate exceeds the implicit tax rate on tax-exempt bonds, the after-tax borrowing rate of these corporations is considerably below that of state and local governments. Furthermore, private investors receive additional subsidies through such mechanisms as the investment tax credit and accelerated depreciation, although the effect of these subsidies is somewhat offset by the fact that private net revenues from investment projects are subject to tax, while state and local revenues are not (Kopcke and Kimball, 1979, p. 21).

The result is that, even with the tax-exempt provision, the public sector must pay more to finance an investment than the private sector. This has led to recommendations to provide a more direct and deeper subsidy for state and local government borrowing—measures that are discussed in chapter 8.

Reduced Demand for Municipal Debt

Higher relative interest rates on tax-exempt securities have been necessary to attract investors in the face of declines in demand (or, more properly, a slower rate of growth of demand) that have resulted from:

☐ Changes in tax laws, including the reduction in the

maximum tax rate from 70 percent to 50 percent, the creation of tax-exempt All Savers Certificates, and the expansion of tax-deferred Individual Retirement Accounts, that have created competitive alternatives to tax-exempt bonds.

☐ Changes in behavior by institutional investors, especially the growth of leasing by commercial banks, that have reduced their demand for tax-exempt bonds. The recently enacted corporate minimum tax is estimated to raise the interest rate on municipal bonds by 60 basis points (Trent, 1982).

☐ Reduced confidence in the fiscal strength of many states and local governments, following the well-published brushes with bankruptcy of New York City, Cleveland, and other cities, and revenue and tax rate limitations that weaken a state or local government's ability to repay.

☐ The rapid expansion of tax-exempt issues for nontraditional purposes such as industrial development, pollution control, and residential mortgages.

☐ Massive increases in new federal borrowing as a share of net annual savings.

The combination of sharply reduced demand for tax-exempt bonds and a rapid expansion in the volume of these bonds has driven up their interest rates relative to taxable issues and squeezed many weaker jurisdictions out of the bond market altogether.[8]

To an unknown extent, some of these measures will encourage an increase in household savings, now running at an annual rate of about $200 billion or about 5 percent of GNP. Yet they will also divert savings from traditional investments such as savings accounts and municipal bonds. Household savings will be the primary source of revenues for municipal bond sales because institutional investors have cut back on the purchase of new tax-exempt issues. Commercial banks and casualty insurance companies have traditionally purchased three-quarters of new state and local issues. However, changes in the corporate tax code have led to a sharp decline in the amount of income that these corporations need to shelter by buying tax-exempt securities. Expansion of leasing activities and high rates of interest have reduced the attraction of municipal bonds. In 1981, commercial banks purchased only 18 percent of net new municipal issues, and casualty insurance companies only 7 percent. Some of this decline is a cyclical pattern, but much of it represents a permanent change (Trent, 1982; Viscount, 1982; Peterson, 1982).

However, state and local governments will have to compete with the federal government for households' investments. Although the two types of securities are not strictly competing—the interest on federal securities is taxable—the huge increase in federal demands for funds will crowd out all other types of debt instruments and drive up interest rates. In 1970, federal securities absorbed less than 10 percent of the new loanable funds available. In 1979, the federal government required only 18 percent of the capital markets' new resources. In 1982, that requirement will have more than doubled. For the first time in peacetime, net federal borrowing is now greater than the net increase in household savings. Deficits are projected to increase significantly during the next three years, and the total federal debt is projected to grow by 50 percent between 1981 and 1985. State and local debt rose by a third of the total increase in Treasury and federal agency debt in 1981. The tax cuts are intended to encourage increased economic growth in the long run, but in the short run the resulting deficit is driving up interest rates. In competing with the federal government to sell securities, Henry Kaufman of Solomon Brothers has said, "state and local governments cannot win."

One of the reasons that they cannot win is that the fiscal crises experienced by New York City, in 1975, and by several other cities since have shaken investor confidence. City governments and statutory agencies demonstrated that they were not always willing to use revenues to pay off bond holders before meeting other obligations. Credit ratings have been downgraded by Moody's and Standard and Poors, and, by some estimates, all municipal bonds must now pay a risk premium in the form of higher rates.[9] Reduced rating can raise borrowing costs by as much as two percentage points. In the past two years, six states have lost their favored AAA status.

Confusion in the Bond Market

There are more than 1.5 million different tax-exempt issues outstanding—sold by state and local governments, special districts, public authorities, and industrial development agencies. Some issues were for billions of dollars, others for less than one million. The issues are purchased by individuals (either directly or through pooled funds), by banks, insurance companies, and even by nonfinancial corporations. Some issues are in the hands of a single investor. Others are held by hundreds of investors.

The growing complexity of municipal bond issues and the proliferation of special authorities with bonding power have also clearly worried investors who are unfamiliar with the type of issue or with the issuing agency. A single bond issue might have thirty different maturities with twelve coupon interest rates. More than 50,000 political entities have debt outstanding. There are no organized exchanges for municipal securities as there are for corporate securities. States and localities do not have to conform to disclosure requirements required by federal law for corporate securities. They use private bond-rating agencies to assess the quality of bonds on behalf of investors—Moody's, Standard and Poors, and Fitch's.[10] Sixty percent (by dollar volume) of municipal debt is rated, but only 30 percent of all issues (Ingram and Copeland, 1982). The "chaotic" municipal security market, as it was labeled by the Joint Economic Committee (1981), increases the costs of public finance.

These changes in the tax-exempt market will not be reversed as inflation abates or when the economy recovers. The structure of the market has been permanently changed (Peterson, 1982) and could only be reversed if the massive tax reforms of 1981 were undone—if personal income taxes were raised, and if other tax-exempt instruments were abolished—events that are highly unlikely. The public sector will continue to face a hostile bond market in the foreseeable future, and must design its capital investment strategy accordingly. Some of the proposed federal initiatives, such as direct interest subsidy payments for a "taxable" public bond, that could reduce public borrowing costs are discussed in chapter 8.

Reducing the Cost of Debt Finance

The ability of state governments to affect the cost of borrowing is very limited. The major factors that have driven up the cost of interest are beyond the control of state governments, and are, in major part, the result of federal actions. Significant changes in federal tax law, in the level of federal deficits, and in monetary policy would be necessary to reduce the cost of long-term state and local debt.

Even where state actions have contributed to the problem—such as through the proliferation of tax-exempt issues for purposes that are not clearly public—it is difficult for any single state to have any real effect on the tax-exempt bond market. It would also earn a reputation for being anti-business or anti-development. A state that abolished its industrial development agencies would have little impact on the overall volume of IDBs flooding the market and would arouse the ire of its business community and many local officials. Again, federal action would be required.

Perhaps the most significant step that a state government can take is to carefully review the purposes for which it is issuing debt. Is the investment one that really requires public funds, or is it more properly the responsibility of a private firm? Is the investment necessary, or are there cheaper and more efficient ways of meeting public service needs? By cutting back on the public debt for industrial subsidies, not-for-profit private health care facilities or private pollution control projects, a state could use its debt for vitally needed *public* investment projects. States (and localities) should also review their use of business tax incentives. Although no accurate data are available—only a handful of states keep records of tax expenditures—evidence suggests that more than $1 billion in potential annual tax revenues is not collected because of abatements, exemptions, and tax credits (Kieschnick, 1981). These potential revenues could be used to finance direct public capital investments.

The next significant step is for the state to provide for adequate revenue sources to back the bond issue and to cover the operation and maintenance of the facility, an issue discussed in detail in chapter 3. The selection of appropriate revenue sources will enable the state to protect its fiscal integrity and so guarantee a high quality bond rating. It is also important for states to provide local governments with the taxing powers necessary to back limited obligation and even general obligation bonds. Sales and property tax increment financing could be a valuable mechanism to ensure that those benefiting from a project pay for it and that the facility will be adequately maintained. Taxes levied by special assessment districts can provide a way of recapturing local benefits from projects. As states develop capital investment strategies in conjunction with their localities, some will have to delegate increased taxing and spending powers to ensure a full and cooperative relationship.

The most important actions states can take to assist local governments to issue debt, is to provide fiscal assistance. Intra-state revenue-sharing programs, categorical grants, and state assumption of program responsibilities will probably have to be increased. However, there are ways in which the state can reduce the costs of local government borrowing that do not involve direct financial assistance but which are based on improving the way the market for tax-exempt issues operates. Forbes and Peterson (1978) classified these mechanisms under four headings:

- [ ] state supervision and technical assistance;
- [ ] financial intermediation;
- [ ] grants for local debt service costs;
- [ ] guarantee of local debt.

There are both advantages and disadvantages of increased state intervention in local bond issues. It provides the state with the power to coordinate state and local infrastructure planning, enforce state policies (such as the imposition of user fees), reduce the probability of a local default (which might weaken the state's own rating), and reduce interest costs. However, this power incurs responsibilities. The exercise of power will necessitate refusing certain bond issues or demanding locally unpopular changes in plans and in local policies. It will, inevitably, cause some local resentment at the intrusion of the state, a problem with obvious political implications. But in most states, increased state involvement in local affairs will be the inevitable result of reduced federal involvement and the expansion of state-run block grants. Some of the promising policies in each of the four areas are described below.

Supervision and Technical Assistance

A study by the National Conference of State Legislatures summarizes the activities in specific states:

> North Carolina and New Jersey provide the most extensive service. The North Carolina Local Government Commission provides. . . ."assistance in setting up new budgeting and accounting procedures as well as training sessions to help local officials manage their city's financial affairs." The Commission also has ultimate responsibility for marketing of local debt issues. . . .
>
> In New Jersey the Local Finance Board exercises extensive regulatory power over local finance. The state has established comprehensive budgeting, accounting and financial reporting standards. For example, local governments must be audited annually and must have their operating and capital budgets approved by the state.
>
> But the state also provides extensive technical assistance and training effort through Rutgers University and the Bureau of Financial Management Services (Watson, 1982, p. 12).

The number of states providing the types of supervision and assistance are shown in table 19.

Financial Intermediation

The myriad of small issues sold by local governments and special districts is an expensive way to raise funds. The transaction costs are high, and investors who are unfamiliar with the issuing agency may require a risk premium, which drives up interest rates. These costs can be reduced through a state agency or authority acting as an intermediary between local governments and the credit market. There are two ways in which this intermediation can occur. The state can issue bonds and use the proceeds either to purchase local government obligations (bond

Table 19

How States Supervise and Assist Local Debt Management

Activity	Number States
State supervises or collects data on local government debt issues	41
— Collects and disseminates data	24
— Maintains data file	22
— Prescribes contents of official statements	14
— Reviews local bond issue	19
— Approves local bond issue	9
— Helps market local bond issue	9
— Other	12
State provides technical assistance to local government in debt management	32
— Helps with official statements	14
— Provides data to	
Issuers for use in official statements	14
Bond rating agencies	16
Underwriters and dealers	18
Prospective investors	11
— Helps evaluate bids	7
— Issues bulletins, pamphlets, manuals	12
— Conducts seminars and conferences	12
— Other	7

Source: National Conference of State Legislatures and Municipal Finance Officers Association, *Watching and Counting,* October, 1977, p. 9.

banks) or make loans directly for specified programs (loan programs).

Bond Banks. Six states have set up bond banks— Maine, New Hampshire, Vermont, Puerto Rico, Alaska, and North Dakota. Only in North Dakota does the state place its full faith and credit behind the bonds it issues to purchase local debt. In the others, the state bonds are, essentially, backed by the credit worthiness of the local governments whose issues are purchased. In Maine, the state issues are also backed by a lien on state grants-in-aid, and, as a last resort, the state's moral obligation. The NCSL study describes the advantages of the Maine bond bank:

Savings are realized on underwriting and marketing costs because of the larger issues and better credit rating of the bond bank. . . . Katzman (1980) estimates that transactions (costs) savings amount to 1.4 percent of the locality's bond issue. . . . The most significant savings realized by the bond bank is from the improved credit rating and thereby lowered interest payments. The bank has a credit rating of Aa, just one step lower than the Aaa rating for the state's own general obligation bonds. The majority of localities selling to the bank have Baa ratings or no ratings at all. . . . To date the Maine Bond Bank has issued $300,000,000 in bonds. It purchases 90 percent of all local government issues in the state (Watson, 1982, p. 20).

Opposition to bond banks typically comes from investment bankers and bond counselors who fear loss of business. Bond banks would not be effective in all states. Empirical research (Jarrett and Hicks, 1977; Katzman, 1979 and 1980) suggests that it is a program that can be successful in reducing costs where:

- [ ] there are many small localities and special districts with bonding power; and
- [ ] the credit rating of many of these local jurisdictions is weak; and
- [ ] the state's credit rating is Aa or Aaa.

Loan Programs. Several states have loan programs that provide local governments with low-cost loans for specified purposes. The programs are usually financed through the issue of state general obligation bonds, but some western and mountain states provide loans to energy boom-town communities from funds built up from severance tax revenues (discussed in detail in chapter 6). In thirteen states, loan programs have been set up to lend money to localities for sewer and water projects. Most of the loans were for the 25 percent match on federally funded sewer and water projects (recent federal budget cuts have reduced the federal share to 55 percent [from 75 percent], which will probably lead to increases in the share that localities can borrow from the states).

An interesting example is provided by the Texas Water Development Fund that lends to localities to build water supply facilities (but not distribution systems). The rate at which a municipality can borrow is one-half of a percentage point above the weighted average of the past three issues of the fund—but with a constitutional upper limit of 6 percent, a constraint that is currently precluding any lending. The upper bonding limit of the fund is $600 million.

Colorado has lent nearly $60 million to localities at 5 percent to finance 50 percent of the construction costs of new water supply facilities. Pennsylvania was recently authorized by the voters to issue $220 million in state bonds for the repair and construction of water supply systems. The interest rate will be equal to the rate paid on state bonds.

Grants for Debt Service

These programs earmark state grants to localities for debt service—in some cases, state monies are sent directly to bondholders. This earmarking of grants is intended to reduce the risk of the bond issue and so reduce the interest rate at which it can be marketed. The most successful example is the New Jersey Qualified Bond Program (Peterson and Miller, 1981; Watson, 1982). This program combines strict supervision of bond issues with direct payments to bondholders. For example, Newark agreed to raise its water rates by 50 percent to receive the board's approval for a water system bond issue. Under this program, localities can earmark state urban aid grants or grants from the state's Property Tax Replacement Program as a guarantee of repayment to the bondholder, without passing through the local budget process. "Purchasers of debt issues are thus guaranteed first access to the locality's most stable revenue source, greatly reducing the risk of default" (Peterson and Miller, 1981, p. 63). The localities must agree to replace these earmarked revenues from local revenues to ensure "maintenance of effort" in meeting local public service demands. The program has allowed Newark and Jersey City to improve their rating on qualified issues from Baa to A, a saving of nearly 200 basis points on each issue. This type of program is particularly effective where a state has a much stronger rating than many of its major cities and if it has an intra-state grant-in-aid program.

State Guarantees of Local Debt

There are various approaches to using the credit worthiness of the state to guarantee local bond issues so that they can enjoy lower interest rates. New Hampshire places it full faith and credit behind local bond issues for school construction and pollution control projects. Michigan guarantees school bonds not with its full faith and credit but with an agreement to sell general obligation bonds to pay interest and principal on local bonds nearing default. Minnesota has a similar program. However, if used excessively, these programs can jeopardize the state's own credit rating.

Conclusions

What all the programs discussed in this final section rest on is the ability of a state government to regulate local bond issues and fiscal practices more efficiently than would be done by local governments acting individually. If the state acts only as a rubber stamp on local issues, then it will lower its own bond rating and lose the ability to cut the cost of local borrowing. The examples briefly discussed here suggest that state and local governments can act together to the mutual benefit of both. But it is not a partnership that is achieved without cost.

Many of the programs also involve states issuing more debt. Although voter approval of bond issues has now climbed back over 50 percent (from only 33 percent in 1975), selling a bond issue is not an easy task. Political leaders must explain the consequences of cutting capital spending. In 1982, in spite of a worsening economic and fiscal environment, Missouri voters broke with tradition and approved a $600 million general obligation bond issue for an array of capital projects including soil conservation, road repair, state university facilities, and the rehabilitation of state office buildings. Governor Bond led a strong campaign to depict the consequences of continued under-investment in public works. Bond issues cannot be sold as a free good. Taxes will have to be raised to service the debt, and trying to shield this from voters allows opponents an easy way to undermine the education process.

Unfortunately, there is little hard evidence to prove that an investment in infrastructure may be a more powerful inducement for economic development than a business tax cut. But the link can be drawn, and, in several states, the business community has been a powerful advocate for bond issues for public capital investment.

CHAPTER V NOTES

1. The U.S. Census Bureau defines public debt as "long-term credit obligations of the government and its agencies, and all interest-bearing, short-term (i.e., repayable within one year) credit obligations" (cited in Gold, 1981, p. 12). These include "judgments, mortgages, and revenue bonds as well as general obligation bonds, notes and interest-bearing warrants. On the other hand, non-interest-bearing, short-term obligations, interfund obligations, amounts owed in a trust or agency capacity, advances and contingent loans from other governments, and rights of individuals to benefits from employee-retirement funds are not included" (cited in Gold, 1981, p. 12). Short-term debt instruments such as tax anticipation notes are designed to bridge gaps between inflows of revenues—which may be received quarterly or even at the end of the fiscal year—and necessary expenditures. However, for the purpose of discussing the finance of long-term public investment, we can exclude short-term debt.

2. Alaska, Delaware, Hawaii, Kentucky, Minnesota, Nebraska, New York, Oregon, Rhode Island, Washington, and Wyoming.

3. Arkansas, Idaho, Indiana, Iowa, Mississippi, Missouri, North Carolina, Ohio, and Utah.

4. Connecticut, Hawaii, Louisiana, Maryland, Massachusetts, Michigan, North Carolina, Oregon, Pennsylvania, Tennessee, Washington, Wisconsin, and West Virginia.

5. Arkansas, Arizona, Colorado, Indiana, Iowa, Kansas, Nebraska, South Dakota, and Wyoming.

6. The single exception is the arbitrage bond issue solely for the purpose of investing in higher yielding taxable securities.

7. The real interest rate is the difference between market rates and the rate of inflation.

8. For a concise summary of these changes, see Herships and Karvelis, 1981.

9. Municipal investors are concerned with four types of risk: default,

marketability, maturity, and business cycle risk. Default risk measures the probability that principal and/or interest will not be paid when due. Marketability risk concerns the ability to convert securities into cash before maturity. Maturity risk relates to the expected rate of interest prevailing at the maturity date, and business cycle risk concerns changes in the overall level of yields from general economic fluctuation (see Ingram and Copeland, 1982).

10. Municipalities must pay for this service, and provide the rating agencies with information on their fiscal position that is not necessarily available to the public (see John Petersen, *The Rating Game*, The Twentieth Century Fund, New York, 1974).

CHAPTER VI

Financing Large-Scale Resource Development Projects

BETWEEN 1967 AND 1970, the population of the city of Gillette, Wyoming, doubled as a result of intensive oil and gas exploration in surrounding areas. During the past decade, the population tripled, this time as a result of coal development (Leistritz and Maki, 1981). Very rapid development in small communities can place extreme demands for the rapid expansion of infrastructure and public services, often beyond the fiscal or institutional capacity of the local community and even the state. Boom towns usually grow as the result of massive private investments in the development of natural resources, such as Rock Springs, Wyoming, and Parachute, Colorado. But they may also arise because of the construction of port facilities, such as Valdez in Alaska, or from the building of military facilities.

Without a carefully designed strategy to deal with boom towns, severe problems can occur. A major study by the Massachusetts Institute of Technology of the impacts of large-scale energy developments in western and mountain states identified eight types of problems (see table 20), including social disruption, shortages of goods, poor quality public services, environmental degradation, and acute fiscal problems. At the same time, corporations attempting to develop the project may be hampered by lack of skilled workers, escalating construction costs because the necessary transportation and communications facilities are not in place, and by expensive lawsuits from environmental groups and local residents. Disruption and

Table 20

Components of the Boomtown Problem

Boomtowns represents different problems to different groups. There are positive aspects of energy development, but even some of those have negative side effects. Our case studies and the literature on boomtowns reveal eight components of the boomtown problem:

1. Social Disruption. Energy development causes sudden changes in the population mix and patterns of everyday life. These in turn cause social problems and social conflicts.[3] Rates of alcoholism, drug abuse, mental illness, divorce and juvenile deliquency increase. While many of these problems are experienced by newcomers unaccustomed to their living conditions, long time residents are the ones most affected by the disruption. For example, long time residents are more likely to become alcoholics or suffer from mental illness than newcomers.

2. Public Service Needs. Americans have come to expect certain basic public services such as roads, water, schools, police and fire protection, and social welfare assistance. During rapid growth these services are often overburdened, or unavailable to some groups. In addition, public services which a town did not provide before may be necessary to support energy development or to cope with its side effects (i.e., counseling may be needed for those suffering from community disruption). New residents sometimes expect more or different services than long time residents. Tax rates must often increase to cover the cost of providing new or expanded services. The lead time needed to design and build new facilities means that the

costs are borne by those who live in the area before the boomtown population has actually arrived.

3. Shortage of Private Goods and Services. During a boom the private market rarely keeps pace with the demand for goods and services, especially housing. In some cases, housing shortfalls can restrict energy development: one hundred families recently found no housing when transferred to an oil boomtown and had to be transferred back to their previous positions.[4]

4. Inflation. Excess demand triggers inflation in prices, wages and rents. While price increases are welcomed by the storeowner whose costs usually do not rise as quickly as revenues, and increased housing prices are a blessing to the landlord, inflation is particularly harmful to the senior citizen and others on fixed incomes who cannot take advantage of rising wages. High construction wages, combined with a general labor shortage, cause other wages to rise. This can hurt an agricultural economy (though agricultural workers benefit from higher wages if their employers do not go out of business). Increased costs can also affect provision of public services. Two boomtowns had to increase salaries 40% in order to hold experienced employees. Increased costs for building materials raise municipal costs just when public facilities need to be expanded.[5]

5. Revenue Shortfalls. Even though growth expands sales and property tax bases, revenues increase more slowly than costs in the short run. Despite a 19% increase in sales tax revenue, one coal boomtown has already increased property tax *rates* several times; even with a 68% increase in its local sales tax revenue, an oil boomtown finds itself short of operating funds.[6] These revenue shortfalls are due to (i) delays between the time development begins and the time the locality realizes either property or sales tax revenue; (ii) delays in raising capital for constructing and improving public facilities; (iii) capital needs beyond local government's legal bonding capacity; (iv) location of high tax-yielding properties outside the communities hosting the newcomers and the resulting public costs.

6. Resources Lost to Other Uses. Industry and its workers are notably consumptive of three resources needed by the agricultural economy; water, land and labor. As new industries use efficient collection techniques and cities exercise eminent domain over water rights, less is available for agriculture. In some states' energy development regions, groundwater use is unregulated by state permits. Increased consumption by energy development may mean water shortages for cities and agricultural producers drawing from the same aquifer.

Easily irrigated land near stream beds is particularly valuable to agriculture but it is also valuable to energy developers because, for example, coal is nearer the surface. When strip-mining removes land from agricultural production (for at least ten years in most places), local food processing industries fail. Even in oil boomtowns, where less land is disturbed, agricultural producers face a shortage of inexpensive labor, since high drilling salaries attract unskilled and semi-skilled farm workers.[7]

7. Aesthetic Deterioration. Boomtown development sacrifices amenity to economy and ease of construction. Trailer courts are laid out without paving or landscaping; commercial establishments are built of sheet metal and often located in unsightly strips along major roads. Many people mention aesthetic deterioration as a problem, particularly if they considered the area attractive before the boom. The size of new developments causes part of the aesthetic problem. Many new neighborhoods, in which trees and shrubs have not had a chance to grow and which look barren, dwarf established parts of town.

8. Fundamental Change. An important cost of boomtown development has nothing to do with conventional indicators of stress or inadequacy, since it results from change itself rather than from what the town changes to. The original residents of a boomtown chose their community—or chose to remain—because it was the best place for them (certainly the best they could afford). When development occurs, the appearance, social structure, friendship patterns, style of life, and nearly everything else about their community changes, and the community that supported them simply disappears. The injury such disappearance causes is only partly mitigated if the "new" town is clean and orderly.

Source: Susskind, Lawrence and Michael O'Hare, *Managing the Social and Economic Impacts of Energy Development,* Massachusetts Institute of Technology, 1977

delays occur not because residents and developers have divergent goals—indeed, they often share a long-run desire to encourage growth—but because there are few effective institutional mechanisms to resolve disputes and to determine how to finance and develop the necessary public infrastructure and services.

Boom town problems are the most obvious manifestations of problems that affect many jurisdictions—that is, private development projects, from shopping centers to factory branch plants, that are large, relative to the community in which they are located, and that create a permanent and widespread change in the local economic base and in the demand for public facilities and services. Although the discussion in this chapter focuses on energy-related boom towns, the types of policies discussed could easily be applied to communities dealing with growth induced by a private recreation development, a new airport, or a housing complex.

There are several costs that must be considered in developing a financing strategy to deal with large-scale developments.

☐ *Initial infrastructure costs.* Before a new project can start to function, new roads or other transportation systems will have to be developed, and the capacity of existing facilities will have to be expanded. Water supply and waste treatment systems will be needed. Utility hook-ups for the new populations must be provided. And the residential facilities—schools, libraries, and police and fire stations—must be built. These investments require front-end financing.

☐ *New Public Services.* After the initial building phase, the city or county must operate at a greatly expanded level—more police officers, teachers, and sanitation workers, and more money to be spent on infrastructure maintenance.

☐ *The "Bust Phase."* For most large-scale developments, the initial boom is followed by a decline in employment as construction labor leaves and as the project adjusts to its permanent operating level of employment. In some instances, the project may prove uneconomical and be closed completely. In other instances, the project may experience strong, cyclical fluctuations in activity as a result of changes in the world price of minerals or of recessions and expansions in the national economy. Because small communities do not have diversified economies, they are particularly vulnerable.

☐ *Compensating those harmed by development.* Some long-time residents will find their means of livelihood destroyed or their lifestyles so changed that they move away. Some will be compensated—at least in part—by increases in property values or by expansions of their businesses. A consistent compensation policy minimizes expensive lawsuits, protests, and other tactics employed by individuals and groups seeking relief.

State governments can provide a framework within which these costs can be allocated equitably and efficiently between public and private agencies and between the state and local governments. This chapter describes this framework. The first section discusses how to assess the social and economic impacts of a proposed project. The second section describes the overall management of state programs to deal with large-scale projects. In the third section, alternative revenue sources to finance necessary public capital investments are analyzed. Programs to provide state aid to impacted communities and individuals are outlined in the fourth section. The final section summarizes the lessons from these programs that are applicable to state public investment strategies. Much of the discussion is based upon an analysis of the experiences of western and mountain states with large-scale energy projects.[1] These lessons should prove invaluable to all states as they design complex but necessary programs to ensure that major developments are compatible with their long-run economic and social goals.

Project Impact Assessment

The purpose of project impact assessment is to develop the information necessary for the state and local governments to ensure that development and growth occur in an orderly manner, that adverse effects are minimized, that those harmed are compensated, and that the costs of building public facilities and providing services can be covered through tax revenues and through cost sharing with the developer, where that is appropriate. The assessment must be able to predict economic, demographic, social, and environmental changes related to the project and relate these changes to resource needs and the requirements for public facilities and services.

The ability to predict the social, economic, and fiscal impacts of a large-scale development project is neither easy nor cheap, and is subject to considerable uncertainty. Between deciding in the spring of 1980 to undertake a massive oil shale project near Parachute, Colorado, and terminating the project in the spring of 1982, Exxon's estimates of its own costs had risen from between $2 and $3 billion to nearly $6 billion (*Fortune*, 5/13/82, p. 61).

Companies typically undertake detailed analyses of the socioeconomic impacts as well as engineering and marketing studies. However, it is difficult for state and local governments to rely upon the data supplied by the prospective developer, however, well-conducted. The research may be biased. Certainly, relying only on the developers' own estimates will arouse suspicion among those opposed to the development. In the case of Parachute, Exxon provided the state with the funds, with no strings, to develop its own impact-assessment study, a practice which has been followed by several companies in other states. A consulting contract of more than one quarter of a million dollars to assess the full impacts of a large project is not unusual. The development of a replicable "assessment model" that can be used on many different projects would reduce the costs of future impact assessments. Skimping on the development phase of the assessment model will lead to high costs later. Colorado, which has developed such a model under the auspices of a "Cumulative Impacts Task Force" (composed of representatives from state agencies, local governments, and industry), has found that the model tracks impacts accurately and can be an invaluable tool in negotiating fiscal mitigation strategies.

Effective impact assessment can meet two problems

that can undermine the process: (1) inadequate notice of the proposed project from the developer; and (2) inadequate access to data and information. Sufficient notice is no problem where the development will occur on federal or state lands. But where a private company already owns land and the necessary development rights, a massive project could occur with very little notice to local residents. One approach to ensure adequate notice is to require all major projects (defined by type of project, size of investment, or level of projected employment) to apply for a development permit from a state agency. Three states— North Dakota, Wyoming, and Montana—have developed effective mechanisms to ensure early notification and permitting processes that ensure that costs are negotiated with the private developer. Utah provides an example of a much less formal approach.

Impact assessment should not be confused with cost-benefit analysis although it should borrow many of the same analytic methods. The purpose is to lay out, as accurately as possible, the likely impacts of a project—both positive and negative—and how these costs and benefits will be shared among individuals, jurisdictions, and institutions. This information can then be used during the negotiation process that determines who pays for the additional costs and how the payments are made.

Obtaining project data is a problem that cannot easily be solved. Assessing a project's impacts requires highly technical data from the developer and also detailed financial data, both of which are usually confidential. Disclosure could lead to widespread financial harm to the corporation and its stockholders and could even jeopardize the viability of the project. In a project where passions run high, there is a very real danger that information could be leaked by parties anxious to block the project or impose their own conditions. A state that gains a reputation for indiscrete use of data will discourage private developers. At the same time, if excessive restrictions are placed upon what type of data can be released, then the results of the impact analysis cannot be verified by analysts employed by interested parties. If published findings are not believed, then interested parties may resort to costly and delaying legal battles.

113

North Dakota[2]

In North Dakota, a facility siting act was passed in 1975 which gives the Public Service Commission (PSC) authority to develop an inventory of potential sites for energy conversion plants and transmission lines (Lu, 1977). Utility companies—under the PSC's jurisdiction—are not restricted to these sites, but the PSC must evaluate any alternate locations chosen outside the existing inventory. Their evaluation is financed through an application fee of 0.05 percent of project cost—a user fee that ensures PSC of the resources necessary to conduct the study. The PSC has the power to override local regulations if a plan is approved. The process requires public hearings. The PSC has so far used its authority to either accept or reject utility siting proposals; it has not conditioned approval on the utility's commitment to provide infrastructure or other forms of impact assistance, such as front-end financing of public investments.

Project impact information is gathered and provided to communities by Interindustry Technical Assistance Teams (ITAT) made up of experts from the coal industry. ITAT uses computer models to provide employment projections through 1990, publishes biannual reports containing updates on development impacts, and works with state and local planners to suggest impact mitigation strategies.

Wyoming[3]

In Wyoming, the Industrial Development Information and Siting Act of 1975 created a council within the Governor's Office. All projects with a construction cost in excess of $50 million (as adjusted for inflation since 1975—the current level is $90 million) and all energy conversion projects in excess of a specific capacity must apply for a permit to the council (Foster, 1977).

The permit may impose conditions on developers, including requiring the provision of infrastructure. Information supplied to communities through the siting act is supplemented by technical assistance from thirteen advisory state agencies.

The siting act sets the stage for negotiations between

local communities and developers. Communities' zoning powers give them some bargaining power against developers and help them allocate their infrastructure resources efficiently. Moreover, local planning powers were reinforced by the Land Use Planning Act of 1975. This act mandated local planning of economic development and allocated up to $20,000 for this purpose to each county for a two-year period. The State also has extensive environmental protection legislation that requires detailed consideration of project impacts.

Montana[4]

Montana has the most complete system for assessing and controlling the impacts of coal (and, more recently, oil and gas) development projects. Most of the legislation, passed in the early 1970s and including the Major Facility Siting Act, was designed to deal with coal-related boom activities. Water pollution controls, air quality standards, and solid and hazardous waste management laws are directed to health and environmental concerns. In addition, special tax laws have been enacted. The Major Facility Siting Act requires anyone proposing energy generating and conversion plants and associated facilities above specified capacities to obtain a Certificate of Environmental Compatibility and Public Need before construction can proceed. The review and approval process is fully financed by a filing fee and sets time limits on the permitting process. The Department of Natural Resources has twenty-two months to study a proposal and to complete an Environmental Impact Statement. The Board of Natural Resources (citizens appointed by the governor) then has up to eleven months to conduct a hearing and issue a decision on the application. The board holds public hearings and, prior to issuing a certificate, must determine the basis of the need for the facility. The board also determines whether the facility represents the minimum adverse environmental impact allowed by available technology and looks into the nature of various alternatives.

When granting the certificate, the board has the right to impose conditions on the construction, operation, or maintenance of the facility. Since the enactment of the siting law in 1973, the board has exercised its power to impose conditions in every certification.[5] The act also requires developers to submit, annually, long-range plans for the construction and operation of covered facilities proposed to be constructed within the state in the ensuing ten years.

Utah[6]

Utah's only law relating to energy development has been on the books for about a year. The Natural Resources Development Act stresses cooperation and has no enforcement provisions. It requires only that a developer submit socioeconomic assessments and mitigation plans to local governments and the Community and Economic Development Department (CEDD), but approval is not required. Since an estimated 80 percent of development in Utah occurs on federal land, the most important weapon state and local officials can deploy is their ability to hold up water permits and building permits. In a few cases, communities have also been able to use conditional zoning to force developers to provide housing for project employees.

In summary, impact assessments can help a local community anticipate future problems, and, therefore, develop ways of resolving some of the problems identified in table 20, even if there are no state funds. But assessments coupled with resources and statutory powers to enforce negotiation between developer and community can be much more useful and equitable.

Managing Large-Scale Project Development

Local communities are rarely able to deal, unaided, with the consequence of large-scale development. First, the impacts of individual projects often cross jurisdictional boundaries. For example, coal development in Campbell County, Wyoming, has occurred outside the city of Gillette although the population expansion has occurred within the city. The county faces little difficulty in financing expanded public services through mineral tax revenues. The city, which does not control these revenues, has inadequate resources and faces a much more difficult task. The recent passage of the Wyoming Joint Powers Act attempts to reconcile these interjurisdictional problems by allowing combinations of local governments and districts to apply for loans to address growth problems.

Second, few small communities have the expertise and the resources to negotiate on equal terms with the developer. Many have only a few, low-paid public employees and may have only a part-time manager or mayor. They are unable to assess anticipated impacts and to prepare a detailed strategy for negotiating with the private developer.

Third, many large-scale resource development projects occur on federally or state-owned lands or on land in which some of the development rights (or rights-of-way) must be secured from a state or federal agency. Because it would deal with many different projects, a statewide agency could compile a wealth of experience in designing agreements that could not be duplicated at the local level.

Fourth, counties and municipalities are essentially creatures of the state and usually require state legislation in order to change the local tax base or to issue debt. Consequently, it is important that large-scale developments be coordinated statewide in order to ensure that projects are consistent with each other and do not incur unnecessary social or economic costs.

Development management should, therefore, be coordinated at the state level—even if the coordinating agency may decide, for any specific project, that no state actions are necessary and allow all negotiations to be conducted between local jurisdictions and the developer.

The Goals of a State Project Management Mechanism

The purpose of a state project management mechanism is to negotiate the allocation of costs among private and public agencies. It is *not* to bleed the project for the maximum tax revenues or to block projects that may be technically incompatible with environmental regulations. It certainly is *not* to add another layer of bureaucracy in what is already, all too often, a very tangled process. In fact, the mechanism, if it is correctly designed and if its procedures are adhered to, should streamline the permitting process for viable projects and allow the development to proceed with less fear of sudden changes in state and local taxes or regulations. The negotiation process should ensure that projects do not lead to sudden local fiscal crises requiring a state bail-out, that fluctuating employment does not require sudden changes in state taxes, and that harmful environmental impacts are reduced. It would enable local governments to anticipate with greater accuracy their capital construction and operating expenditures. Local residents will be more confident that their interests have been considered, that the best possible bargain with the private developer has been struck, and that those suffering most from the project have been compensated. Development interests have been served, while, at the same time, considerations of equity and environmental quality will have been included. Most important, the effort to balance interests is clearly visible and recognized by all concerned parties.

The state strategy should, therefore, include the following functions

1. Estimate the effects of development—social, economic, and fiscal—before final decisions are made by the developer and the community.
2. Coordinate the permitting process among the relevant public agencies.
3. Ensure the representation of all interested parties in discussing and negotiating the development.
4. Encourage the negotiation of an agreement between interested parties in a timely and equitable fashion.
5. Coordinate administrative and legislative changes necessary to carry out the project.
6. Monitor the development of the project to ensure compliance with the terms of the negotiation and to respond to any unforeseen problems and developments.

The Negotiating Process

For a negotiation to be successful, it is necessary to spell out several particulars in advance:

1. The topics to be negotiated;
2. Who can participate in the process;

3. An arbitration vs. mediation procedure that can be used if negotiations reach a deadlock;
4. A fixed time period within which the agreement must be reached.

Typically, the agreement will identify specific financial and administrative responsibilities and allocate them among the private developer and state and local governments. It may even carry conditional responsibilities that change the terms of the agreement if prescribed circumstances change during the course of the project. For example, the tax rate on mineral extraction may be reviewed if market prices of the mineral increase above, or fall below, predetermined benchmark levels.

Establishing a monetary value on all aspects of the negotiations may be neither possible nor efficient. For example, environmentalists may resist direct payment from a mining company in compensation for the spoiling of a natural area—questioning to whom the payment is made and what it is used for. They may prefer the company to purchase other, unspoiled, private land and turn it over to the state or county to be used as a wilderness area in perpetuity. The context of negotiations should be sufficiently broad to allow this type of exchange.

A negotiating process may seem alien to those firmly schooled in traditional cost-benefit project analysis. They might argue that if enough resources are devoted to developing economic, environmental, and other models, then the efficient and equitable allocation of costs can be determined. This is not the case. Such models and available data are notoriously unreliable. Many of the affected parties—both the local community and the developers—do not feel bound by outcomes computed by technicians, and thus challenge findings in courts or through the political process. The result may be a breakdown in the planning process and much greater uncertainty, which deters investment and multiplies distrust. Traditional cost-benefit analyses do not allow for the subtle trade-offs that must be made in a large-scale and complex development project.

Experience with negotiated investment strategies is limited. The complex negotiations that preceded the construction of the Alaska pipeline are probably an example of what not to do. Agreements (between the federal government, the consortium of ARCO, Exxon, and Standard, and the state) were not based upon clearly established principles and were reopened several times. It is probably impossible to completely prevent renegotiation. Once a company has begun investing in a project, the negotiating power of the state or locality is strengthened. If the company strikes a bonanza, the state's position is further reinforced. More recent procedures established in Montana, Wyoming, and Colorado have been much more successful. They illustrate that major developers are aware of the benefits of a negotiated development and are prepared to act on this awareness. The process does not involve the surrender of any of the legal rights of those participating and, therefore, does not preclude any subsequent lawsuits, even though the fact of participation and public hearing in the negotiating process may deter later legal battles.

Determining who should participate in the negotiation is not easy. Which environmental group should negotiate on behalf of local citizens concerned about the possible environmental impacts? Will its negotiation strategy reflect the interests and preferences of the local community? How many state and national environmental concerns should be involved? How should delegates or representatives to the negotiation process be selected? What interest groups should be recognized? How should the results—preliminary and final—be communicated to the developers and to the local community?

There are no simple answers to these complex questions, although judicial solutions do provide examples. In the *United States v. Allegheny-Ludlum Industries* (5th Circuit, 1975), the trial judge created a panel of thirty-six people to represent six groups of plaintiffs. As experience with negotiations increases, we shall learn what approaches work most effectively. A brief description of a few examples of negotiated projects illustrates both the processes used and the outcomes that can be achieved.

North Dakota

Local governments use information from the Interindustry Technical Assistance Teams to manage local growth patterns through zoning ordinances and through negotiations with utilities and mining companies. A good example is provided by the development of Coal Creek Station, North Dakota's largest mine mouth power plant. Total investment for the plant transmission lines and coal mine was about $1.2 billion. The construction work force at the plant in 1979 totaled 4,620 people while the permanent operating workforce during this period equaled 800. Coal Creek Station utilities contributed $40,000 for local public works, improved roads around their development site, organized a Citizen's Advisory Board to hear citizen concerns, and stimulated housing development before the start of the project by assembling a large tract of land and selling it to a residential developer.

The state also made contributions. Between 1975 and 1980, the Coal Impact Office made grants and loans to Washburn and Underwood, the two towns most directly impacted by Coal Creek, including, $747,000 in grants to the county for roads; $330,000 in grants to Washburn; $225,000 in grants to Underwood for expansion of water and sewer systems; $807,000 in grants to the county for additional police and fire protection services; and $570,000 in grants to Washburn and $520,000 in grants and $35,000 in loans to Underwood, all for schools.

State officials and industry representatives seemed satisfied with North Dakota's procedures for dealing with energy developments. Surveys of local residents during and after the Coal Creek construction revealed that, "almost all the persons interviewed indicated that the construction of the Coal Creek Station had a very positive effect on the community."[7]

Wyoming

The State's Industrial Siting Council successfully negotiated with the Missouri Basin Power Association to minimize the impact of the construction of the Laramie River power generating station in Wheatland. The Association agreed to provide up-front money for infrastructure development as well as planning assistance to the affected local governments.

Wyoming also provides examples where private developers have acted to coordinate their plans with local communities. In the development area known as the Overthrust Belt, which runs through Evanston, Kemmerer, and other small towns, the oil and gas developers formed the Overthrust Industrial Association (OIA). The OIA acted as local impact coordinator, providing interest-free loans to communities until a tax base was established, giving grants, constructing community improvements, providing impact projections, and helping with maintenance, such as snow plowing and road repair. In a more extreme example, ARCO, the developer of the Black Thunder Project, built the whole town of Wright. ARCO apparently had two reasons for this. Under the state's siting law, ARCO would have had to pay infrastructure costs in any case, and Gillette, the nearest town, was over forty miles away. A primary reason for ARCO's decision was that it was cheaper to build a town at the project site than to lose twenty work days each winter when snow storms close the roads. ARCO currently employs between 400 and 500 people at the development site.

Colorado[8]

Carbondale, once one of the towns most impacted by energy development projects, experienced most of the typical infrastructure problems associated with a development boom in the mid-1970s. Mid-Continent Coke and Coal Company responded to the town's request for assistance by providing a $10,000 planning grant, forming a housing corporation, and providing private buses to transport employees to and from the mines. On the whole, however, companies were reluctant to assume a substantial share of the front-end costs associated with capital intensive improvement projects. Localities had little bargaining power because they were worried that developers would go elsewhere and because they had little political leverage with the state legislature.

More recently, negotiations have proved much more successful both from the perspective of the local communities, which have benefited from private funding for public infrastructure, and for the developers, who have been able to proceed with the project in a more orderly fashion. Exxon provided nearly all the infrastructure needed for the Colony Oil Shale Project near Parachute, Colorado. When Exxon and Tosco started the project, the population of Parachute was about 400. When Exxon abruptly closed the project in the spring of 1982, it had to fire 2,100 employees working at the site. To support its workforce, Exxon started to build the town of Battlement Mesa, complete with stores, apartment complexes, a large recreational vehicle park, subdivisions (with $90,000 homes), and a partially completed recreation center costing more than $3 million. Exxon also loaned the local school district $12 million to finance front-end costs until a local tax base could be created.

Officials at the State Energy Impact Office attribute the greater cooperation between private developers and state and local officials in recent projects to several factors.

1. Possibly the most important reason is the formation in the state of an urban constituency interested in the social and environmental aspects of development projects. Major corporations cannot afford to antagonize large segments of the population.

2. Studies by these corporations have concluded that the increased productivity and lower project costs make it economical for the developer to support the provision of local public facilities and services.

3. State policies have encouraged cooperation. The state and local governments have shown in the past that they can stall projects by holding up building permits. The state has facilitated cooperation by conducting a voluntary permitting procedure, the Colorado Joint Review Process, that involves all interested parties at an early date. All parties are currently satisfied that cooperation is mutually beneficial.

Colorado uses several mechanisms to coordinate its energy policies. In 1975, Governor Lamm created the Energy Policy Council to develop state energy policy and to make recommendations on specific problems. Members of the council include the executive directors of the State Departments of Natural Resources, Local Affairs, Agriculture, and Health, as well as members of the governor's staff. The Western Slope Energy Impact Committee

(WSEIC), consisting of local citizens and staff from county planning agencies, was formed to provide the Energy Policy Council with local information. The Economic Impact Office works with developers to provide communities with periodic impact projections. Area-wide program coordination is carried out through thirteen councils of government. Each council is made up of the county commissioners of the area and municipal officials. The councils serve as the intermediaries between local government and the state and are the agencies through which reviews of many state and local funds pass.

The state and private developers have helped local governments with front-end financing and with community-specific impact projections. Communities have also used their own power to affect local development. Local governments, in addition to using their zoning powers, have switched to metered water to manage use and have tried to attract retail establishments into growing communities.

Revenues to Finance
Public Infrastructure
for Large-Scale Developments

Most large-scale resource projects have involved the extraction of minerals, principally gas, oil, coal, and oil shale. States have relied heavily on revenues generated by severance taxes to finance needed public facilities and services. Corporate income taxes, lease revenues, property taxes, sales taxes, and shared federal lease revenues have also been used. To secure up-front funds, some states and localities have required prepayment of taxes by the developers or have received low-interest-rate loans from the development corporations for public infrastructure construction. To ensure that revenues do not suddenly cease when the project is completed (or abandoned), some states pay a portion of their revenues into trust funds, funds whose principal or earnings can be used to diversify the local economic base during the bust phase. All states with trust funds protect the principal but allow the interest to be spent for more general purposes. During times of high inflation, spending all of the interest allows the principal to decline in value. Ideally, only the *real* interest would be spent, with the remaining interest returned directly to the fund to maintain its real value.

Before discussing specific state revenue-raising policies, some general remarks about the advantages and disadvantages of alternative taxes should be made.

☐ *Severance Taxes.* These are easy to collect and extremely popular among state governments. To the extent that project impacts are related to the volume of minerals extracted (for example, the volume of strip-mined coal is strongly related to the level of damages to the environment), a severance tax is an appropriate source of revenues. The principal impact of the tax is to reduce the value of mineral-bearing land and, therefore, the value of leases. A high severance tax (which is usually collected by the state) will reduce revenues from leases on state lands and from property taxes on private lands (which are locally collected). The tax is paid by a corporation regardless of its profitability (unlike the corporate income tax). This may deter the development of marginal mineral industries (such as coal in Alaska). It may also make local industries much more vulnerable to sudden declines in world mineral prices because they must continue to pay the tax even if the operation is no longer profitable. Severance taxes, because of their nature, are useful to finance infrastructure development costs and also, when deposited in trust funds, can finance the bust phase of the project and compensate those harmed.

☐ *Corporate Income Taxes.* Many states already have a corporate income tax which is paid by mineral extraction companies. The tax has the advantage of relating tax payments to the ability to pay. It will tend to spur a faster rate of mineral extraction. One disadvantage of the tax is that many mining companies are multinational, as well as multistate, and can allocate their income among different locations, placing it beyond the taxing-reach of individual states. It is also difficult to use the corporate income tax to recapture massive windfall benefits that may accrue from sudden increases in mineral prices. Revenues from this tax are appropriate in funding ongoing state services to the project community.

☐ *Lease Revenues.* Many resource projects are located on state or federal land. Leases are usually sold determined in public auction. Revenues may provide the state with front-end revenues which may be shared with local communities. The federal government rebates to the state about 50 percent of the revenues from the sale of leases on federal land a share of which must be distributed to impacted counties according to a formula. These revenues can provide a useful source of front-end funds.

☐ *Property Taxes.* These are the primary source of revenues to finance local public services. Their principal disadvantage in rapidly growing communities is that revenues do not adjust as rapidly as public service demands. Also, much of the property on which the development occurs may lie outside the town or county that experiences most of the impacts.

☐ *Personal Income Taxes.* This tax can provide the state with revenues that track the boom-bust cycle, providing that temporary construction workers pay taxes within the state (company withholding can ensure that they do).

☐ *Other Taxes.* Sales taxes, especially if shared with localities, provide a revenue source that also tracks the boom-bust cycle. Both sales and property tax increments can be used to back local bond issues if permitted by the state.

An ideal revenue base to meet the needs of very rapid development would include all these tax elements. Excessive reliance on one tax source will not adequately meet the multitude of demands and the change in these demands over time. It may also distort the pattern of development.

The techniques used by mineral-rich states vary greatly. Montana has the most extensive financing system. Texas and Arizona have virtually no program to deal with fiscal mitigation and, therefore, no dedicated funds. The different approaches provide some insight into the alternative policies and their effectiveness.

Montana

Since 1975, Montana has imposed a coal severance tax set at 30 percent of the mine-mouth price, although the actual tax rate is about 22 percent due to numerous available deductions and credits for local property taxes paid by coal companies. In fiscal year 1981, total coal severance tax revenues were about $70.4 million. Coal tax revenues must by law go into separate funds or accounts. The division of coal tax revenues into these accounts is presented in table 21.

The permanent coal trust fund is designed to compensate future generations for the present exploitation of a nonrenewable resource. The interest earned on this fund is appropriated, but the principal can only be expended by a three-quarters vote of each house of the legislature. In 1981, all of the interest was appropriated for highway maintenance and construction. At the end of fiscal year 1981, the permanent fund totaled $75.2 million.

Wyoming

Wyoming's approach to development projects is similar to Montana's. Wyoming's coal severance tax is set at 10.5 percent, and has a net effective rate of about 9.5 percent after deductions and credits. The tax revenue goes into five funds, as shown in table 22.

The permanent fund, at the end of 1981, contained about $203 million. The principal of this fund is constitutionally protected; the interest goes into the state general fund. The Community Impact Fund and the Coal Impact Fund are both used for impact mitigation, with the latter fund reserved especially for communities impacted by coal mining and related activities. Both funds are administered by the Farm Loan Board (FLB). The Wyoming Community Development Authority (WCDA) has the power to float up to $100 million in tax-exempt bonds. The money the WCDA raises is loaned to local communities to finance public facilities and public housing. The grants given by the FLB are intended to be used as pledges to repay WCDA

Table 21
Distribution of Montana's Coal Tax Revenues

Fund or Account	% of Total Coal Tax Revenues
Permanent Coal Trust Fund	50%
State General Fund	19
Education Trust Fund	10
Local Impact Area Fund	8.75
Public School Equalization	5
Alternative Energy Research Fund	2.25
Park Acquisition & Cultural Trust Fund	2.5
Renewable Resource Development	0.625
Water Development	0.625
State Library Fund	0.5
Counties—Local Land Use Planning	0.5
Conservation Districts	0.25
TOTAL	100%

Table 22
Distribution of Wyoming's Coal Tax Revenues

Fund or Account	Tax Rate	% of Total Coal Tax Revenues
Wyoming Permanent Fund	2%	21%
Community Impact Fund	2	21
General Fund	2	21
Coal Impact Fund	1.5	16
Water Development Account	2	21
TOTAL	approx 9.5%	100%

loans. In the past, FLB grants have been restricted for use only on highways, road or street improvements, and water or sewer projects because these were the most pressing needs. Recent FLB grants have gone for other capital facility needs such as hospitals or mental health centers as well as for equipment such as fire engines.

In 1981, the state imposed a two percent oil and gas severance tax. Half of the proceeds will be distributed directly to cities, town and counties.

North Dakota

North Dakota has created an impact-assistance fund from the revenues generated by a coal severance tax. The current tax rate stands at $1.01 per short ton of coal; the rate is indexed to inflation, so that the tax rate increases $.01 for each four-point rise in the Wholesale Price Index. In 1981, the coal tax generated revenues of almost $17 million. The allocation of coal severance tax revenue is determined by statute and is shown in table 23.

The statute allocates coal tax revenues among different state funds and among different political subdivisions within the county where the coal was mined. If the tipple (loading facility) of a mine is within fifteen miles of a non-producing county, both counties receive a share of coal revenues according to a formula specified by statute.

A Coal Conversion Privilege Tax is applied to electrical generating plants and coal conversion (e.g., gasification) plants. This tax rate currently stands at $.25 per kilowatt-hour of power generation. In 1981, more than $3.4 million in conversion tax revenues were collected. These tax revenues are also allocated by statute as is shown in table 24.

Reflecting the growth in oil and gas developments in North Dakota, the state set up an Oil and Gas Impact Program in 1981. A direct appropriation of $10 million was made from the general fund to finance grants to local political subdivisions. The program parallels the Coal Impact Program. Local subdivisions submit applications for assistance for specific projects. The director meets with applicants and makes final funding decisions, based upon the degree of negative impacts resulting from oil and gas activity, the extent to which the proposed project will reduce those impacts, and the fiscal need of the applicant (which includes the local tax effort, the level of community cooperation, and the amount received by the applicant from the Oil and Gas Production Tax and other energy-related revenues).

Colorado

Colorado enacted legislation creating a coal severance tax, at a rate of $0.63 per ton (inflation adjusted) for surface coal and $0.31 per ton for underground coal, and setting up a Local Government Impact Assistance Fund in 1977. The state also has severance taxes on other minerals and oil and gas. In 1980, severance tax revenues totaled $31 million, $11 million from coal, $8 million from oil and gas, and the rest from other minerals. Since 1981, 50 percent of the severance tax revenues are distributed to the Local Government Severance Tax Fund, which funnels these revenues back to counties and municipalities affected by energy development loans. The other 50 percent is deposited in the Severance Tax Trust Fund to be utilized for the future consequences of the depletion of mineral resources and water project development. The interest from investments of the Severance Tax Fund is deposited in the general fund. As required by the federal government, Colorado distributes a share of federal mineral royalties to counties and school districts.

Utah

Since 1979, Utah has operated a Community Impact Account which has received 32.5 percent of the state's share of mineral lease revenues. It distributed over $20 million to energy-impacted communities. In 1982, the legislature, meeting in special session, created the Permanent Community Impact Fund and placed in it $35 million from oil shale bonus funds, 70 percent of all future bonus bid funds, and 32.5 percent of the state's share of mineral lease revenues. The fund will operate as a revolving loan fund to aid energy-impacted communities in making public capital investments.

Assistance to Localities

Since responsibilities for public services and public capital investments are shared between state and local governments, the funds to meet fiscal mitigation needs must also be shared. The type of aid provided, how it is allocated, and the conditions attached will significantly affect the ability of local jurisdictions to meet their needs. Again, examples from energy states provide useful lessons.

North Dakota

The county's share of the coal conversion tax revenues and the coal severance tax revenues are subdivided

Table 23
Distribution of North Dakota's Coal Tax Revenues

Fund or Account	Tax Rate in $		% of Total Tax Revenues	
State General Fund	$ .30		30%	
State Trust Fund	.15		15	
Coal Impact Fund	.35		35	
County Where Coal Mined	.20		20	
— County General Fund		$.08		8
— Incorporated Municipalities (based on relative population)		.06		6
— School Districts (based on number of pupils)		.06		6
TOTAL	$1.00	$ .20	100%	20%

Table 24
Distribution of North Dakota's Conversion Privilege Tax

Fund or Account	Tax Rate ($s/kw-hr)		% of Total Tax Revenues	
State General Fund	.1625		65%	
County Where Facility Located	.0875		35	
— County General Fund		.035		14
— Incorporated Municipalities		.0263		10
— School Districts		.0263		10
TOTAL	.25/ kw-hr	.0875/ kw-hr	100%	35

according to the statutory formula shown in tables 23 and 24: 40 percent of the county's share goes into the county general fund, 30 percent goes to incorporated municipalities within the county, and 30 percent goes to school districts within the county.

The funds created to receive coal severance and conversion tax revenues are very similar to those created by Montana and Wyoming. The coal impact fund is administered by the Energy Development Impact Office (EDIO), which makes funds available ($6.5 million in 1981) to local political subdivisions to help them to pay for public services and facilities. The EDIO evaluates each request for impact assistance according to the specific needs of the locality.[9]

Aside from Coal Trust Fund loans and grants from the EDIO, communities can also draw upon the coal tax revenues that are returned directly to them. There are no restrictions as to how counties, municipalities, and school districts may make use of this money. Communities receive additional state aid through general revenue sharing; North Dakota, however, does not have a coal or mineral property tax that communities can draw on. Federal mineral leasing royalty payments go only into a school trust fund.

Wyoming

In addition to loans from the Wyoming Community Development Authority, impacted communities also receive state aid from two other sets of funds and accounts. Wyoming, with nearly 50 percent of its land under federal ownership, receives substantial federal mineral leasing royalties (over $100 million in 1981). While many states just add these royalty payments to the general funds, Wyoming puts 50 percent of the royalty *bonus* payments into the Government Royalty Impact Assistance (GRIA) account administered by the Farm Loan Board (FLB). The state share of the royalty payments themselves are placed into nine different funds. About 14 percent of the returned royalties are given directly to local political subdivisions for local expenditures including roads and schools. A further 16 percent of the state's royalty funds are earmarked for additional special impact assistance.

These two programs provide considerable assistance to communities. In fiscal year 1980, $60 million in coal tax revenues were collected and something less than $100 million in federal royalty payments were received, while about $28 million in impact assistance was given out. Almost all of this money is administered by the FLB. Most communities are eligible only for loans and conditional grants because they will eventually be able to pay back current deficits with future increases in tax revenues. FLB loans and conditional grants are forgiven if the anticipated economic development and tax revenues do not occur. Communities receive additional funds from the state through intra-state revenue sharing programs, and, since

1973, counties can—by referendum—impose an optional 1 percent sales tax. Seven counties, including those containing Gillette and Rock Springs, had enacted the optional tax by the beginning of 1977.

In 1981, the state created a program that allocates funds from the state's share of the increase in sales and use taxes collected in counties where a facility is under construction with a permit issued from the siting council. Only those counties levying the 1 percent local option are eligible to receive the funds from the state. The funds are distributed within the impact counties to the jurisdictions according to each's share of the sale tax distribution. The siting council may determine that an adjoining county is impacted and provide for a ratio-of-impact formula to be used to distribute the funds between the counties. The revenues from the oil and gas severance tax, enacted in 1981, will provide an estimated $250–$300 million in aid to localities by fiscal year 1986. Finally, Wyoming addresses the problem of interjurisdictional cooperation within the state by allowing local governments to form Joint Powers Boards, which act strictly as administrative agencies, consolidating the borrowing and lending authorities of local governments.

Montana

The local impact fund in Montana is administered by the State Coal Board and is used for local impact assistance. After passing through an application process, loans and grants are given to local governments for governmental services and facilities needed as a direct consequence of coal development. Since its inception, the State Coal Board has awarded 122 local impact assistance grants, totaling $40.2 million (twenty-three to the Colstrip area). Most of the funds were used for public facilities and capital equipment.

Unlike most states, Montana does not have a revenue sharing plan with local governments, but portions of specific taxes or fees are distributed to local governments on a formula basis and often for specific uses. For instance, Montana returns to all counties a portion of state liquor revenues. The School Foundation program helps to fund maintenance and operating costs but not capital expenditures in local school districts. Counties collect a gross proceeds tax on the production of coal within their boundaries. Montana has not yet resolved the issue of interjurisdictional cooperation when the benefits of development are enjoyed in one jursidiction but most of the costs are incurred in another.

Colorado

Of the monies placed in Colorado's Local Government Fund (50 percent of severance tax revenues), 15 percent are automatically distributed to impacted jurisdictions in proportion to the number of employees of each mine who reside in each county's unincorporated area or in each municipality. The remaining 85 percent is distributed at the discretion of the executive director of the Department of Local Affairs (DLA) to impacted communities for planning, construction, and maintenance of public facilities and for the provision of public services. The tax resources and tax effort of the communities are considered in awarding the grants.

The Local Government Severance Tax Fund and the Local Government Mineral Impact Fund (which receives federal leasing royalties) are coordinated by the Division of Commerce and Development (C and D) in the DLA. C and D has encouraged the formation of local impact assistance teams, composed of local and county officials, industry representatives, and the general public, to assign priorities for local funding requests.

The Oil Share Trust Fund, established in 1974, receives funds from federal oil shale leases. The entire proceeds are spent in the four impacted counties for public facilities and services—appropriated by the state legislature. Overall, the extent of assistance that can be provided by the state is limited because of the state's constitutional prohibition against indebtedness.

Texas[10]

Texas provides an example of development with little assistance to local governments. Typical boom town problems exist in southwest Texas due to oil and lignite development, and in East Texas as a result of lignite mining and the construction of coal-fired power plants.

> While previous oil and natural gas development has taught Texans to protect the physical environment, they have drawn a different conclusion about the social environment: cities with serious rapid growth problems during the oil boom days have survived (Stinson, 1977, p. I-5).

Perhaps as a result, Texas has no state financial impact assistance program, very little monitoring or predicting of community-specific impacts, and also very little local government power to zone or otherwise negotiate with developers. Public and private utilities were recently denied the power of eminent domain for mineral development. Counties in Texas do not have the power to pass ordinances or zone, and most towns have very limited powers to annex or pass ordinances. As a result, developers have given communities little help in providing infrastructure.

Although many communities have suffered, some have come up with innovative means of controlling economic development even without state assistance. For example, the town of Pearsall charges utility rates that are 50 percent higher outside the city than inside. The town of Dilley gives no guarantees that utility connections will ever be provided outside the city limits. However, owners of land outside the city limits who petition for annexation often receive priority treatment. The town of Mt. Pleasant took the unusual step of seeking cooperation with the county and school district so that they could share infrastructure (e.g., paving equipment, buildings, fire protection).

Conclusions and Recommendations

Our limited experience with large-scale project development suggests several lessons for all states and localities that can be applied to developing an equitable and efficient process for allocating the costs of public facilities and services and for ensuring orderly development.

135

Intergovernmental Responsibilities

The capacity to finance the public facilities and services necessary for large-scale projects involves federal, state, and local governments as well as the private developers. Indeed, many of the project sites span federal, state, and private lands. New initiatives will be necessary at all three levels of government if a fully successful strategy is to be constructed.

Federal Government. The federal government has so far seemed unwilling to get involved in discussions about infrastructure responsibilities. It has not provided significant amounts of impact mitigation assistance (all state officials interviewed agreed that the amount of federal aid was so small as to have no effect on socioeconomic impacts) nor has it explicitly considered impact mitigation planning in its decision on whether to lease particular portions of federal land. In addition, the federal government may constrain state policy choices for future negotiations between state and local governments and developers by unilaterally making the decision to lease a particular portion of federal land.

At present, the federal government does reimburse a part of the lease revenues to the originating state, which proved important to Colorado in its negotiations with Exxon for financing the Colony development. But the federal lease payments are often lower than the price paid on comparable private land, and the amount returned to the states is not large—usually much less than is returned from state-owned land. The development of a coherent state investment strategy could be strengthened by a more consistent leasing approach and more active communications between state and federal officials.

State Responsibility. The state government has three levels of responsibility. First, it must promulgate a consistent overall policy toward large-scale resource development projects. This policy must involve a clearly delineated siting policy, a consistent environmental framework, the fiscal framework for financing these projects, and the clear delineation of the role of the state relative to local governments. Second, it must provide a framework within which local governments can negotiate with private developers, the subjects that are negotiable, and the broad procedures that must be followed during a negotiation process. Third, it must provide real resources—both technical and fiscal—to local government units that provide them with the tools and power to negotiate effectively. These must include a detailed estimate of possible impacts and the power to improve the necessary tax or to issue the necessary debt to pay for the "public" side of the negotiated strategy.

In order for a state to negotiate or to legislate an effective capital investment infrastructure strategy, it must first be clear about its own values and about its goals. These values do differ among states. Residents of Montana are worried about the environmental and social damage resulting from development projects. Montana thus retains tight control of the development process and charges a high severance tax. Utah and Texas apparently want more energy development and so put very few restrictions on the development process. Each state must examine its collective values in order to make an intelligent cost-benefit analysis of what it stands to gain by pursuing economic development in general or the continuation of a specific project in particular. Some of the important factors to be considered by the state include the costs and the benefits that have been discussed in this chapter.

The state should also evaluate specific possible outcomes. It should consider, for example, how much it values the revenues or economic activity induced by the project. It must judge the value of the private developer's commitment to contribute further to the state's economic development in the future. For example, a commitment by the developer to build processing, refining, or manufacturing facilities in the state would be very valuable. And the state must assess the value of the developer's provision of infrastructure.

The goals of the state should include, at a minimum, that it not suffer a net loss by adopting a particular development policy. If costs and benefits are defined broadly enough, and weighed by the state's own collective values, this goal could be interpreted as choosing the development policy that would maximize the net benefits to the state.

Local Governments. The responsibility for negotiating the development of a large-scale project will rest, ultimately, with the local government units—townships or counties—that house most of those directly affected by the project. However, the responsibility for setting the overall framework within which these negotiations are conducted, for conducting evaluations and analyses of options, and for providing fiscal resources, must rest with the states. In states where local capacity is clearly not sufficient to undertake sophisticated negotiations, the state must be prepared to take primary responsibility in conjunction with representatives from the impacted communities.

Having recognized that a favorable settlement on infrastructure responsibilities must be weighed against other possible bargain outcomes, it next becomes necessary to consider the fiscal tools that the state and its communities can bring to the negotiating table.

Financing Tools

States and localities can use a variety of fiscal and regulatory tools to finance project evaluation, public services, and infrastructure necessary for large-scale project development.

Impact Assessment and Evaluation. A state office to undertake impact assessment and to coordinate the negotiation process is an essential component of an effective strategy. This could be located within the governor's office or within a line agency with jurisdiction over resource development. This operation can be successfully financed by a "project application fee" as in North Dakota and Montana. It could also be financed through the prepayment of a portion of taxes from the project.

Infrastructure Development. Where most of the development of a new town can be attributed to a single project, a large part of the costs can be imposed directly upon the developer. Other mechanisms that do not involve the issue of public debt include: (1) the prepayment of taxes; (2) proceeds from lease sales on state land;[11] (3) federal "revenue sharing" from the proceeds of lease sales on federal land; or (4) a development fee that reflects the cost of extending roads and sewer and water services to the project. If public debt is used, then limited obligation bonds can be issued backed by: (1) the increment in sales tax or property tax revenues; (2) the revenues from a severance tax or corporate profits tax imposed on the development companies; (3) the revenues from special district taxes—a multi-county area affected by the development could be granted special district status and allowed to impose a supplementary sales or property tax.

Public Services and Mitigation Payments. These should be financed by the additional tax revenues that are generated by the project, including all additional revenues from the development companies, local businesses, and residents. If their costs are not covered then the project is being implicitly subsidized by other taxpayers in the state, which may lead to reduced development elsewhere. It will be important to monitor public sector revenues and costs attributable to the project to minimize such subsidies. Some states—notably Alaska, Texas, and Louisiana—have chosen to use revenues from oil and gas production to subsidize all other sectors of economic activity. These subsidies cannot be sustained in the long-run which may lead to traumatic adjustment problems when oil and gas tax revenues decline. There is little evidence to suggest that subsidies are necessary to "prime" development, and therefore, wherever possible, the development of new industries or regions should proceed on a pay-as-you-go basis.

The Bust Phase. Most mineral-rich states have recognized that booms cannot continue indefinitely. Resource prices fluctuate widely over time, and easily recoverable resources are eventually depleted. Most states set aside some of their annual revenues in trust funds that have three major purposes: (1) to invest in new industry that will diversify the local economy; (2) to provide state and local revenues that will continue as severance taxes and other resource-related revenues decline; and (3) to repair the environmental damages incurred as a result of resource development. Although this is a prudent course, the typical level of investment in trust funds—with the exception of Montana and Wyoming—is insufficient to provide very much cushion. In most states, revenues are used overwhelmingly to fund state operations—a procedure that promises severe fiscal pain in the future.

Large-scale development is not necessarily harmful nor is it necessarily an unmixed blessing. A carefully designed procedure to negotiate an equitable and efficient allocation of the costs among state and local governments and private developers and the appropriate deployment of fiscal policies can lead to a harmonious and mutually beneficial development that will build and sustain healthy local economies. This would involve a much greater state commitment to impact assessment and negotiated development then has yet been made in most states.

CHAPTER VI NOTES

1. The information presented in this chapter relies heavily on the excellent study conducted by the Laboratory of Architecture and Planning, Massachusetts Institute of Technology, which examined policies and practices of fiscal mitigation in mountain states. This study is summarized in Susskind and O'Hare (1977), but see also Foster (1977) and Gilmore and Duff (1975) for a discussion of Wyoming; Sanderson (1977), Texas; Monaco (1977), Colorado; Lu (1977), North Dakota. Overviews of policy issues are contained in Freudenberg (1976); Sanderson and O'Hare (1977); O'Hare (1977); Brody (1977). The discussion of specific state programs is based upon a paper prepared by Klaus Kolb of the Kennedy School of Government, under the supervision of Professor Herman Leonard.

2. Information for North Dakota is summarized from Lu (1977); from correspondence from James Luptak, Deputy Director, Energy Development Impact Office, State of North Dakota, and Darrell Ohlhauser, Acting Federal Aid Coordinator, State of North Dakota; and from a telephone interview with August Keller, Director, Energy Development Impact Office, State of North Dakota.

3. Information for Wyoming is summarized from Foster (1977); from a telephone interview with Warren White, Director, State Planning Coordinator's Office, State of Wyoming; and from correspondence from W. Don Nelson, Governor's Office, State of Wyoming.

4. Information for Montana is summarized from a telephone interview and correspondence with Lee Berry, Director, Department of Natural Resources; Nancy Owen, Don MacIntyre, Frank Culver, Carol Massman, and Paul Cartwright, Department of Natural Resources,

State of Montana; Mike Shields, Lieutenant Governor's Office; Terry Cohea, Governor's Office; and Carol Ferguson, Department of Commerce.

5. For example, the certificate for the power plants known as Colstrip 3 and 4 contains twenty-six conditions, including:
   1. keeping a 50-day water supply,
   2. building sealed sludge ponds,
   3. placing air quality monitoring equipment on the nearby Northern Cheyenne Indian Reservation, and
   4. providing for the training and employment of members of the Northern Cheyenne Tribe.

6. Information for Utah summarized from telephone interviews and correspondence with Kent Briggs, Executive Assistant, Office of the Governor, State of Utah, and Gary Tomsic, Deputy Director, Department of Community and Economic Development, State of Utah.

7. Gilmore et al., Case Study #1: (1977) Coal Creek Station, McLean County, North Dakota, in Leistritz and Maki, (1977, pp. 70–71).

8. Information for Colorado summarized from Monaco (1977) and telephone interviews and correspondence with Lee White, Director, Office of State Planning and Budgeting, State of Colorado, and Randy Russell, State Energy Impact Office, Department of Local Affairs, State of Colorado. Articles on the closing at the Colony Shale Oil Project in *The New York Times Magazine*, May 3–17, 1982.

9. For example, in the case of a temporary population surge, the EDIO may agree to reimburse 100 percent of the cost of portable classrooms. If the community wants to invest in permanent school additions, the EDIO will pay only a portion of the total expenses. However, the community would also be eligible for loans from the Coal Trust Fund. These loans, administered by the State Land Board, are made at 6 percent interest with ten-to-twenty-year payback periods. The state deducts loan payments from a county's share of coal tax revenues. If a coal mining project ceases, the loan is forgiven.

10. Information for Texas is summarized from Stinson (1977) and from a telephone interview with Mr. Mike Martin, Staff member, Senate Committee on Natural Resources, State of Texas.

11. The use of proceeds from lease sales on state lands is restricted in Wyoming. State lands were granted to the State by the federal government under this Act of Admission. Under the terms of the trust created by this Act, income from these lands can only be used for the betterment of Wyoming's public schools and other public institutions.

CHAPTER VII

Using Tax Incentives to Finance Public Investment[1]

A MORE DRAMATIC ALTERNATIVE for financing public infrastructure than those discussed in preceding chapters is to involve directly the private sector by transferring ownership of a public facility to a private firm in order to take advantages of the tax incentives offered to private investment. Harnessing tax incentives for public purposes is hardly new. To finance low-income housing and industrial development, the public sector has devised programs explicitly using tax shelters available to private investors. Over the past two decades, special provisions have been added to the Federal Tax Code to encourage private investment or expenditure that meets policy goals. These provisions do not reduce the cost to government of achieving its objectives but shift the cost from direct on-budget expenditures to less-direct tax expenditures. This chapter analyzes how state and local governments can take advantage of federal business tax incentives as a means of cutting the cost of infrastructure finance. It describes the limits to private participation and examines the circumstances under which private funds might be used to finance public facilities[2] that have traditionally been conceived of as purely public in nature.

The mechanisms described in this chapter—lease financing, service contracts, and outright sale—are, essentially, fiscal gimmicks. They do not reduce the actual cost of making the investments. They do not improve the efficiency of public investments through improved planning and management. They simply export a part of the cost of a local project to federal taxpayers. As such, they are fiscally equivalent to a federal categorical grant program that may fund between 10 and 15 percent of a project's cost.

Unfortunately these mechanisms operate less efficiently than federal categorical grants. Part of the federal tax expenditures—the federal corporate income tax revenues foregone—flow not to the municipality but to the private corporation entering into the financial agreement and to the financial institution negotiating the deal. Indeed, not all the tax burden is necessarily shifted across jurisdictional boundaries. For example, a large share of the cash payment by Metromedia to the New York Metropolitan Transit Authority for the purchase of buses reflected not saved federal taxes but reduced state and city corporate income taxes.

However, what may be a relatively inefficient national policy may be attractive to individual states and cities. After all, even if a categorical program provides grants to localities worth less than the cost of raising the necessary federal revenues, localities that do not apply gain nothing. With federal direct grants sharply curtailed and tax-exempt interest rates remaining stubbornly high, tax-incentive financing is worth considering. The appeal of this approach to state and local government is similar to the appeal of Industrial Revenue Bonds (IRBs) since, in both cases, federal tax expenditures fund part of the costs. There is a certain amount of irony here. Through IRB financings, state and local governments assist private, taxable businesses by creating the convenient fiction that tax-exempt IRBs have a fully public character. The same governments, through lease finance arrangements, seek federal tax subsidies for a purely public project by depicting the project as a private, taxpaying entity.

Transferring ownership of a public facility to a private corporation—even if only for tax purposes—raises the issue of why the facility was public to begin with. The belief in the appropriateness of public ownership extends beyond mere cost considerations and cannot be simply answered. In determining whether tax-incentive financing can be applied to a given facility, the following general questions must be answered:

1. If a facility cannot or, it is determined, should not cover its costs (because it is in part a public good or merit good), is it possible to channel public funds to a private facility operator to capture the tax advantages of private ownership? Theoretically, the cost to the public sector of operating the facility might be reduced by the value of tax breaks for private investors.

2. Where the facility possesses the characteristics of a social good, can conversion to private ownership—with appropriate regulations—allow cost savings and, at the same time, meet public use requirements?

3. Where the facility has all the characteristics of a private facility, including the ability to generate surplus revenue, can the outright sale to a private company generate increased revenues to the public sector?

The first section, below, reviews the tax advantages of private ownership in terms of the investment tax credit and accelerated cost recovery and analyzes the legal impediments to the public sector capture of these benefits. The subsequent three sections describe, respectively, how leasing, service contracts, and sale arrangements can be used to reduce the costs of public facility financing. The final section summarizes the main conclusions and outlines the policy concerns that such arrangements may create for state and local governments. The discussions in the three major sections are highly technical. The reader interested in reviewing the policy issues may wish to turn directly to the final section.

The Tax Advantages of Private Ownership

Government use of various business tax subsidies[3] for public purposes faces certain legal restraints. This section briefly describes the structure of federal tax incentives for business investment, outlines the legal impediments to their capture by state and local government, and defines the types of arrangements that can be made between public and private investors.

Federal Tax Incentives for Business Investment

Under present federal tax law, business investment in plant and equipment is subsidized in two ways. The most straightforward is through an investment tax credit, which is a relatively clear and discrete subsidy. A taxpayer making a qualifying capital investment may deduct a percentage of the cost of the investment directly from the taxes due to the federal government under the corporate (or personal) income tax. The Economic Recovery Tax Act of 1981 (ERTA) established the following percentages:

3-year class life equipment	6%
5-year and 15-year class life equipment	10%
Rehabilitation of structures 30 to 39 years old	15%
Rehabilitation of structures at least 40 years old	20%
Rehabilitation of certified historic structures	25%
Investment in research and development in excess of the level of investment in 1980	20%

In addition, ERTA significantly revised the method by which businesses depreciate capital assets by adding an implicit subsidy to what had been ostensibly a system for merely recovering capital costs. Prior to ERTA, businesses were allowed to depreciate an asset by deducting from income a percentage of an asset's capital cost over the asset's presumed economic life, creating, for tax purposes, a series of expenses that conformed with a hypothetical rate at which the asset was used up. Of course, any such series of expenses can only be approximations in practice. The system that preceded ERTA was extremely complex—classifying eligible plant and equipment under 130 categories that could be depreciated over periods between eighteen months and forty years, and, for most assets, allowing a choice of three ways in which depreciation provisions could be apportioned over the life of the asset. However, no allowance was made for varying rates of inflation, since depreciation was against the purchase price of the asset. Therefore, the effective corporate tax rate on capital rose and fell with the inflation rate. This makes it extremely difficult to devise a depreciation schedule that is equitable for all types of assets and for all rates of inflation.[4]

Despite such practical and theoretical difficulties, it is clear that, beyond a certain reasonable range approximating the true or economic depreciation rate, a radical shortening of asset life for tax purposes and allowance for a more rapid write-off in early years, which ERTA has done through the Accelerated Cost Recovery System (ACRS), is an implicit subsidy for capital investment. ACRS groups all assets into one of only four categories that can be depreciated over fifteen years (structures), ten years (investments by public utilities), five years (most equipment) and three years (transportation equipment and machinery previously depreciated over four years or less).

The changes in the rate at which assets can be depreciated for tax purposes are:

Type of Asset	Pre-ERTA Asset Depreciation Range System (ADR)	Post-ERTA Accelerated Cost Recovery System (ACRS)
Short fixed equipment (weighted average)	5 years	3 years
Other equipment (weighted average)	8 years	5 years
Structures	40 years	15 years

ACRS subsidies are harder to quantify than ITC subsidies. The amount of the subsidy is the increased tax forgiveness over that which would be available if only economic depreciation were allowed. Consider a machine costing $1,000 that wears out evenly over ten years. The real value of depreciation is $100 a year for ten years. Under ACRS the equipment will be depreciated over five years at $200 each year. The total value of the depreciation allowances is the same—$1,000. But with ACRS, the benefits are enjoyed sooner and are therefore more valuable. If we discount future depreciation benefits to compute the *present value* of depreciation under the two systems, the implicit subsidy under ACRS becomes apparent. Assuming a 10 percent discount rate and a 46 percent marginal tax bracket, the present value of depreciating over ten years is $283, and the present value of depreciation over five years is $349. Thus, a business that is confident that it can earn at least 10 percent interest on cash in hand would be prepared to pay at least $66 ($349 minus $283) for the ability to use ACRS rather than the true depreciation rate.

Since there is no rigorous way of calculating the true depreciation rate, we can only approximate the extent of the subsidy. The combined effect of both ACRS and ITC allows for negative tax rates under certain conditions. Alan Auerbach (1982), in a recent examination of ACRS, found the following effective tax rates for assets purchased during 1981–94 with inflation at 12 percent:

Life of Asset	Effective Tax Rate
3 years	– 0.85
5 years	– 0.74
15 years	+ 0.15

A negative tax rate is a difficult concept to understand. Essentially it means that the present value of the tax credit and depreciation subsidies is considerably greater than the present value of the expected earnings from the equipment and therefore will create an internal "tax loss" that the corporation can write-off against its earnings on existing capital. Of course, these estimates are based on hypothetical assumptions about the rate of inflation and the earnings from the investment. But these data illustrate an important conclusion: Because of these changes in the tax law, the after-tax cost of capital investments by private corporations—particularly in relatively short-lived equipment—has been substantially reduced. Therefore the potential benefits of private investment relative to public investment have been dramatically increased.

Legal Impediments to Tax Benefit Transfers

In order to capture private sector tax benefits, agreements between state and local governments and private firms must comply with federal tax laws and regulations. There are many legal impediments that must be overcome. The clearest prohibitions in federal law affecting the ability of government to utilize tax benefits arise with respect to so-called "Section 38 property"—which is defined as property eligible for the federal investment tax credit. Since the ITC was clearly and explicitly designed to be an investment stimulus to private business, Section 38 of the tax code excludes from ITC eligibility property used by government. ERTA qualified this general prohibition by excepting expenditures for the rehabilitation of public structures. Property owned by government but used by the private sector may qualify for the ITC under certain conditions.

Since depreciation is (at least in principle) merely a method of recovering capital costs, there has been no attempt to apply Section 38-type restrictions to depreciation claimed under ACRS. Generally, ACRS may be taken by any taxpayer with respect to property held for production of income or in a business or trade. Thus, while government property *per se* is not depreciable since it is typically tax-exempt by definition, property owned by the private sector and leased for profit to a state or local (or even the federal) government may be depreciable. Similarly, property owned by the government and leased to the private sector for business purposes, such as publicy owned industrial space, may be depreciable under certain circumstances.

If tax subsidies more than offset the traditional advantages of tax-exempt financing available to the public sector, government can save money by leasing facilities and equipment that have been built and are owned by the private sector instead of purchasing them. The lease payments by a government to the private owner for an office building, for example, could be less than the public debt service payments that would be necessary had the government itself financed the building through issuing bonds. However, the IRS limits the type of transactions that can be characterized as a lease.

1. The lessor must have a positive cash flow and profit from the lease independent of the tax benefits.

2. The lessee must not have an investment in the lease and must not lend any of the purchase costs to the owner.
3. The lessee must not have the right to purchase the property at less than fair market value.
4. The use of the property at the end of the lease term by a person other than the lessee must be commercially viable.
5. The lessor must have a minimum investment at risk of 20 percent throughout the lease term.

These standards, which apply with most force to equipment,[5] are meant to ensure that a transaction is truly a lease—that is, the private company owns and maintains the property for profit. They were not originally developed to deny government the ability to capture tax subsidies. They were created to inhibit the unlimited transferability of tax shelter benefits between private firms. In the absence of any such standards defining what constitutes a lease, a firm with no taxable income would never acquire a capital asset. Instead it would lease the asset from a firm with a positive taxable income. The lease payments to the high bracket lessor might create a negative, or breakeven, before-tax cash flow, but a positive after-tax cash flow; the lessee could purchase the asset for a nominal sum at the close of the lease.

These so-called true lease requirements reduce the potential benefits that a state or local government can enjoy. However, they can be circumvented under certain conditions.

1. ERTA specifically allowed for certain transactions to be characterized as leases, regardless of the IRS standards discussed above. Such leases, known as "safe-harbor" leases, allow for the unfettered flow of tax benefits between firms with different brackets, the same flow that earlier IRS rulings had attempted to limit. But safe-harbor leasing activity is limited by ERTA to Section 38 property, which excludes property used by government. (ERTA even excluded from safe-harbor provisions Section 38-eligible rehabilitation expenditures leased to government.) The two ex-

ceptions made by ERTA were mass transit rolling stock and equipment used by volunteer fire departments. New York's Metropolitan Transit Authority realized savings of 22 percent on subway cars and 13 percent on buses by entering into a safe-harbor arrangement; similar savings could be expected in non-transit areas, depending on the tax benefits associated with each type of property. Massive federal deficits, however, led to the immediate curtailment and eventual phase-out of safe-harbor benefits under the 1982 Tax Act. For as long as the provisions do remain in effect, small- to medium-sized public transit operators should explore the benefits of safe-harbor leases, particularly, if purchases by a number of operators can be made jointly, providing for economies of scale in the sale of tax benefits to outside investors.

2. Real property is not fully subject to the IRS guidelines. In particular, the lessee may participate in the financing, and the 20 percent minimum interest rule does not apply. This liberalization, combined with ERTA provisions allowing the ITC to be claimed in connection with real property leased to government, allows for quite deep federal tax subsidies with respect to certain government real property.

3. Arrangements may be structured between governments and private corporations that are explicitly not leases, thereby avoiding IRS lease restrictions. One such mechanism is the service contract, discussed later.

4. In the case of a revenue-producing facility, outright sale to the private sector may be the most advantageous policy for the state or local government.

Even if these three loopholes cannot be used for a particular project, a true lease may still be a cost-effective financing option. True lease requirements do not prohibit a private firm from enjoying certain tax benefits from property leased to government. Since ERTA allows the ITC to be taken along with ACRS with respect to rehabilitation expenses associated with property leased to government, it is possible that capital costs incurred by the public sector under a true lease arrangement may be lower than through alternative financing mechanisms, especially for projects involving rehabilitation of existing structures. The local government will have to provide for a positive before-tax cash flow to the investor and for sufficient funds to purchase the asset at a more than nominal price at the close of the lease, unlike a safe-harbor lease. Yet the tax benefits may still be deep enough to offset these additional public sector costs.

Leasing

This section describes how a true lease agreement between a local government and a for-profit corporation can be used to finance a project, and shows the cost savings that are possible. To show more concretely how leasing restrictions may affect the flow of tax benefits, let us examine a simple case of a safe-harbor arrangement. The main effect of the safe-harbor provisions is to replace existing IRS restrictions with far looser guidelines for defining a lease. Most importantly, they eliminate the requirements that the lessor have a positive cash flow during the lease term and that any repurchase be at fair market value. Although the Tax Equity and Financial Responsibility Act of 1982 limits safe-harbor activities, it is worthwhile examining a safe-harbor transaction, because it shows the workings of tax-incentive financing in its simplest form.

Table 25 presents the data from such a hypothetical transaction. In this example, five-year ACRS property owned by a government, which is eligible for the ITC, is sold to a corporation paying a marginal corporate tax rate of 46 percent and then leased back to the government. The transaction involves a down payment plus a note from the lessee (the government) to the lessor (the private owner). The annual lease payment from the lessee to the lessor is exactly equal to the amount due on the note; the two cash payments cancel each other out (columns 6 and 9). The lessor, however, depreciates the property on an accelerated basis over five years and takes a credit in the initial year of 10 percent of cost. These two factors combine to create significant paper losses in the first five years, allowing the lessor to shield other income from taxation at the

Table 25

Sale/Leaseback Example

($ in millions)

Year	ITC	Depreciation Deductions		Tax Savings		Total debt Service
		Actual	Dis-counted	Actual	Dis-counted	
0	10	15.0	15.0	6.9	6.9	—
1		22.0	19.1	10.1	8.8	15.4
2		21.0	15.9	9.7	7.3	15.4
3		21.0	13.8	9.7	6.4	15.4
4		21.0	12.0	9.7	5.5	15.4
5		—	—			15.4
6		—	—			15.4
7		—	—			15.4
8		—	—			15.4
9		—	—			15.4
10		—	—			15.4
11		—	—			15.4
12		—	—			15.4
13		—	—			15.4
14		—	—			15.4
15		—	—			15.4
TOTAL +	10	100.0	75.8	46.1	34.9	230.9

Notes: * Assumptions: Property worth $100 million and eligible for 5-year ACRS treatment is sold by the lessee. The buyer/lessor puts $10 million down and pays the remaining $90 million, plus interest at 15 percent, in 15 equal annual installments. (The lease also runs for 15 years.) Future amounts are discounted at a 15 percent annual rate to yield present values as of year zero. 10% ITC, 46% tax rate.

Columns may not add up to totals due to rounding.

Debt service			Rent Minus Interest Costs (taxable income)		Tax payable	
Interest	Principal	Rental Payments	Actual	Discounted	Actual	Discounted
—	—	—	—	—	—	—
13.5	1.9	15.4	1.9	1.6	.9	.7
13.2	2.2	15.4	2.2	1.6	1.0	.7
12.9	2.5	15.4	2.5	1.6	1.2	.7
12.5	2.9	15.4	2.9	1.6	1.3	.7
12.1	3.3	15.4	3.3	1.6	1.5	.7
11.6	3.8	15.4	3.8	1.6	1.8	.7
11.0	4.4	15.4	4.4	1.6	2.0	.7
10.4	5.0	15.4	5.0	1.6	2.3	.7
9.6	5.8	15.4	5.8	1.6	2.7	.7
8.7	6.7	15.4	6.7	1.6	3.1	.7
7.7	7.7	15.4	7.7	1.6	3.5	.7
6.6	8.8	15.4	8.8	1.6	4.1	.7
5.3	10.1	15.4	10.1	1.6	4.7	.7
3.8	11.6	15.4	11.6	1.6	5.3	.7
2.0	13.4	15.4	13.4	1.6	6.1	.7
140.9	90.0	230.9	90.0	24.7	41.5	11.8

Net tax benefits are equivalent to column 1 (PV of ITC) plus column 5 (PV of depreciation benefits) minus column 13 (PV of tax). Deep tax subsidies in early years more than offset tax due.

Maximum price = 10 + 34.9 − 11.8 = 33.1

Source: Adapted from Mark Willis, "Leasing—A Financial Option for States and Localities," *Federal Reserve Bank of New York Quarterly Review,* March 1982.

46 percent marginal bracket (columns 1–5). However, holding aside the lessor's depreciation deductions, the taxpayer's taxable income grows each year, since his cash rental income is subject over time to smaller interest deductions, which results in net taxable income after the sixth year (columns 10–13). However, these taxes are significantly less in present value terms than the tax savings during the first five years. In fact, the positive difference between the tax savings and tax liability in present value terms defines the amount of equity the investor will be willing to advance. In short, the present value of the ITC plus the present value of tax savings from depreciation minus the present value tax liability from rent income net of interest deductions is the value of the transaction to the private investor and will be the maximum amount that the investor will be willing to pay the government for the right to own the facility for tax purposes.

In this case, the investor should be willing to increase his equity position above $10 million. With each increment in equity, the rate of return drops because of the higher equity contribution and the reduced interest deductions. The investor will continue to add to his equity—in this case up to about $25 million, assuming a typical rate of return—with the offsetting public sector lease costs declining with each increment. The tax benefits provide the private sector lessor the required return to equity to make the transaction possible and profitable.

This transaction shows how IRS restrictions may depress the value of the transaction. In this example, the lessee's net benefit would be about $25 million, since no real economic transactions take place after the initial cash payment. However, if the lessor is required to show a before-tax profit, the lessee's (the government's) $25 million windfall will be reduced because the lessee will have to pay higher lease payments to ensure the lessor's profit. Similarly, the lessee's gain will be reduced by the present value of the repurchase cost in excess of the nominal sum allowed under safe-harbor rules. The IRS guidelines reduce—but may not eliminate—the potential dividend to the local government. But once it is realized that a dividend is legally possible, any number of structures are possible.

☐ *An existing assest may be refinanced.* Certain public facilities may be sold to the private sector and leased back. The financing might be obtained privately or, in some instances, through the issuance of industrial development bonds. In the safe-harbor analysis presented, the maximum amount of the private sector contribution in order to secure tax benefits on $100 million of equipment was judged to be approximately $25 million. In an equivalent non-safe-harbor arrangement, the $25 million in potential tax benefits would not accrue in its entirety as a dividend to government. It will be split three ways: Some must fund IRS profitability tests; a portion must be used to establish a separate "sinking fund" used to repurchase the asset; the balance represents the net fiscal dividend. How much of the maximum $25 million must be used to fund federal compliance depends on a number of factors, including the jurisdiction's estimate of residual value. However, since government retains some ability to depress residual value (through separate ground leases or renewal options at reasonable rates), a substantial portion of free cash is left.

☐ *Sale/Leaseback for renovation.* In the case of a public facility in need of rehabilitation, a sale/leaseback arrangement may provide all or part of the funds that would otherwise be raised publicly. In this case, private monies can fund profitability tests and repurchase, but the balance would be applied toward the rehabilitation of the facility. Since the ITC on real property rehabilitation may be as high as 20 or 25 percent as compared with 10 percent in the equipment safe-harbor analysis, the benefits might be substantial.

☐ *Sale/Leaseback for new construction or acquisition.* Both real property and equipment would qualify for sale leaseback for these purposes.

Service Contracts

A second method by which tax benefits may be put to public use involves service contracts. A service contract is the use by a private vendor of its own equipment to provide a service to a client for a fee. It can be distinguished from a traditional lease (in which the client purchases the right to use an asset for a specified period) and from a lease-purchase arrangement (by which the client acquires the asset over time). But distinguishing a service contract from a lease, while apparently simply a matter of draftmanship, is conceptually difficult and legally hazardous. Perhaps the best way to approach the issue is by examining a relatively clear case of what a service contract is and discussing, later, the limits of the concept.

Copying machines are a fixture in most government offices. "Property used by government" is not Section 38 property, and, therefore, expenditures by the company supplying the government with the copying machines would appear to be ineligible for the federal ITC. However, the Xerox Corporation maintained a service contract agreement with the federal government, through which the federal government purchased copying services for a fee. Xerox claimed the ITC for the cost of the machines positioned in federal offices and defended the action by maintaining that the copying machines in question were not used by government, but rather by the Xerox Corporation, which used them to provide a service to government. The federal government's payments therefore represented a fee for copying services rather than a lease payment on a machine. IRS disagreed. The Court of Claims, in *Xerox Corp. v. the United States,* (I.R.S. Court of Claims, Washington, D.C., 1978) sided with Xerox and the opinion provides some guidelines in terms of what constitutes a legal service contract.

In general, the court agreed that it is not sufficient merely to characterize an agreement as "not a lease" to avoid the "use by government" prohibition. Rather, the actual nature of the owner's possessory interest requires examination on a case-by-case basis. Xerox was able to show, among other things, that it still had responsibility to maintain the equipment and that, since the machines were readily substitutable in the event of mechanical problems, government was not intending to rent a particular machine but rather the benefits associated with any such machine.

In addition, the court put forward, as a second test, the notion that the equipment, to be eligible, should be part of a larger network in which the taxpayer has additional responsibilities. Again, Xerox was able to qualify.

However, two problems arise in attempting to broaden this line of reasoning to make it suitable for wide application. First, the Court of Claims decision, while attempting to set firm standards, inevitably created some ambiguity. Automobiles and trucks, for instance, were specifically mentioned in the decision as equipment not eligible for service contract treatment. Yet an automobile driven by a government employee is not unlike a Xerox machine. Both machine and auto provide a service; both are substitutable; both may be maintained by a vendor; and both may be part of a larger network.

Legal ambiguity does not, of course, necessarily result in prohibition. Depending on the risks the parties are willing to run and the extent to which the equipment user is willing to provide indemnification to the investor, service contracts may be possible for infrastructure or related equipment. In particular, as the notion of infrastructure grows broader through technological change to include sophisticated telecommunications systems and data processing equipment, service contracts on such equipment may be defended more easily than, say, a truck fleet to maintain county roads, if only because analogies to copying machines may be clearer and more direct. However, truck fleets and other traditional infrastructure-related uses ought not to be overlooked by enterprising governments; the fact that the courts have been able to distinguish a particular automobile lease from a particular copying machine service contract does not imply that all arrangements involving automobiles are necessarily leases. Structuring a service contract for vehicles may well involve altered management and control responsibilities, and it will certainly involve adroit legal draftmanship, but such agreements seem theoretically possible.[6]

Not all service contract arrangements, however, are likely to run into severe legal challenges, particularly if the actual use of the equipment in question is arguably private in nature. Resource recovery facilities, for example, may be operated privately or publicly, depending on governmental preferences, market conditions, and other factors. A government may choose to construct and operate such a facility directly, or it may choose to allow for private operation, with a portion of the revenues to the operator to be pledged by government in the form of guaranteed minimum tipping fees. A group of Massachusetts localities have, in fact, recently adopted this approach. It is unlikely that the private owners will be denied the relevant investment and energy credits since they clearly maintain a possessory interest in the facility. From the public's point of view, such privatization is minimized as a political issue if the facility is new, if public sector jobs are therefore not at stake, and if the jurisdiction is relatively indifferent to the issue of public or private control.

In order to determine how cost savings are effected by means of a service contract, we will contrast a service contract with the most likely alternative method of property acquisition—a lease-purchase arrangement.

How a Lease-Purchase Works

A government wishing to acquire a capital asset, but unwilling to incur debt to do so, may enter into an agreement with a private entity, such as a bank, which provides credits for the purchase and then leases the asset with all responsibilities of ownership to the government. Provided the agreement is properly structured, the lease payment by government to the private sector may be treated for tax purposes as though it were a debt service payment, with separate principal and interest components. The tax benefit that accrues to the private lessor (the bank) is the ability to deduct the implied interest portion of government's lease payments from income. In this sense, cost savings to government are effected in the same manner as are all facilities or equipment financed with tax-exempt debt. Lease-purchase agreements typically differ from government debt-financed transactions in terms of security— usually because repayment is conditioned upon annual appropriations rather than through bond convenants— but it is fair for our purposes to conceive of lease-purchase and tax-exempt debt as roughly equivalent in terms of cost, rate, interest, and use of the asset.

How a Service Contract Works

In a service contract arrangement, the private sector provides the financing for the asset, as is the case under a lease-purchase. However, the private firm does not then lease the asset to government. Rather, it maintains ownership and enters into a contract with government to provide a service with the asset in return for a fee. Since government is neither acquiring, paying for, or even legally using the asset, the fee has no principal and interest components. And since the private firm receives no interest, it is afforded no interest deduction and consequently pays tax on the entire fee as ordinary income.

In the absence of depreciation deductions or tax credits, the inability to deduct interest income would drive up the after-tax cost of capital relative to lease-purchase, and the higher cost would be passed along to government in the form of higher fees. However, the private firm will be willing to lower its service contract fees below lease payments if the tax subsidies associated with ownership exceed the tax subsidies stemming from the ability to deduct interest, all else being equal.

Whether the tax breaks associated with private ownership are greater than the interest income deduction on what amounts to public debt depends on: (1) the life of the asset, (2) the type of depreciation and/or credit for which it is eligible, (3) the ability of the private firm to leverage debt for acquisition, (4) the residual value of the asset at the end of the lease, and (5) the investor's tax bracket. We can determine the relative importance of these variables by examining a hypothetical lease-purchase transaction as a base line for comparison. Let us assume the private firm invests $10 million to acquire a piece of equipment that government wishes to use, and that the firm will effectively invest $10 million in cash since IRS regulations prohibit double-dipping—that is, the firm may not incur debt, investing such borrowed funds in tax-exempt securities, in order that it may simultaneously deduct the interest payments on the debt incurred and enjoy tax-exempt interest income.

The equipment is leased to a jurisdiction with provision for purchase for a nominal sum at the close of the lease term. Although the lack of a bond covenant for security may raise the implicit interest rate above that associated with a traditional government debt instrument, let us assume that the timing and amount of lease payments generates a yield equal to a tax-exempt issue—in this case, say, 12 percent.

The data relevant to investor are shown in table 26. Since interest income is deducted, and no tax is paid on

---

Table 26

Hypothetical Lease Purchase Agreement (Cash flow, tax consequences and return on investment (ROI) $10 million lease-purchase 12% implicit interest rate, five years)

($ in thousands)

Initial investment: $10 million

Year	Lease Payment	Interest	Principal	Tax	After-Tax Cash
1	2,774	1,200	1,574	0	2,774
2	2,774	1,011	1,763	0	2,774
3	2,774	800	1,975	0	2,774
4	2,774	563	2,211	0	2,774
5	2,774	297	2,477	0	2,774

After-tax ROI = 12%

Table 27

Hypothetical Service Contract (Cash flow, tax consequences and return on investment (ROI) $10 million service contract 175% d-b; 10% ITC; no residual value; 50% bracket)

($ in thousands)

Initial investment: $10,000

Year	Before Tax Fee Income	Depreciation	Taxable Income	Tax Benefit	ITC	After-Tax Cash
1	2,774	3,500	(726)	334	$1,000	4,108
2	2,774	2,275	499	(230)		2,545
3	2,774	1,479	1,295	(596)		2,178
4	2,774	1,373	1,401	(644)		2,130
5	2,774	1,373	1,401	(644)		2,120

ROI = 11.1%

---

principal repayments, the firm's after-tax cash flow and before-tax cash flow are equivalent, and provide the firm, therefore, with a 12 percent after-tax return on investment (ROI).

One way of comparing a service contract with a lease-purchase is to ask if the same payment by government as a contract fee, will allow for an equivalent or greater internal rate of return to the firm. Table 27 shows that, under a service contract, the firm's ROI drops to 9.5 percent, assuming a $10 million cash investment, 150 percent declining balance depreciation over five years, a 10 percent investment tax credit, and no residual value. Choice of a less-rapid method of cost recovery will depress the return further; a positive residual value will increase the return. Compared with the relatively straight-forward return discernable under a lease-purchase, the firm's after-tax cash flow under a service contract agreement is subject to a variety of influences.

In this case, the fee payment that the public sector would have to generate to provide a 12 percent return on investment would be slightly higher than $2.774 million per year. But what would be the effect of changing certain of our assumptions? Table 28 shows how variations in assumptions alter the private sector ROI, using a lease-purchase agreement yielding 12 percent as a base line in each instance.

The data presented in this table demonstrate that two variables will be decisive in determining whether, and to what extent, financial savings can be expected under a service contract arrangement. First, for short-lived assets, the jurisdiction's estimate of the residual value of the equipment at the end of the service contract is crucial. Under a lease-purchase, government has acquired the asset over the period of the agreement; under a service contract the vendor will typically control the asset at the close of the contract. Private ownership will not be a problem if the asset is obsolete at this point, but ACRS asset-lives are often shorter than assets' useful lives.

A more important consideration, however, is the ability of the private firm to leverage debt—that is, its ability to purchase the asset with the minimum commitment of its own cash. Since interest payments on borrowed funds may not be deducted from income if such funds are used to purchase tax-exempt securities, the firm's investment in a lease-purchase may not be leveraged as in a typical purchase. The ability to deduct interest expenses distorts investment preference in favor of debt over equity, all else being equal; the double-dipping prohibition effectively neutralizes this distortion and converts the lease-purchase investment in equipment or structures into a tax-exempt security. However, no such prohibition exists with respect to a service contract.

Therefore, even if the two tax-benefit scenarios (interest income deductions versus business tax write-offs) have equivalent effects on a given transaction, the ability of the firm to incur debt as a means of reducing equity while also deducting debt service interest payments may create a preference for service contracts. Put another way, the preferential tax treatment of service contracts would allow for a fee payment lower than the corresponding lease-purchase payment, still generating a rate of return satisfactory to private investors. For example, table 28 shows that the private sector ROI increases to 65.2 percent given the following conditions: (1) 90 percent debt financing, (2) private debt financing interest rates 2 percent above the public tax-exempt rate, (3) zero residual value, (4) a three-year asset, and (5) 150 percent declining balance depreciation. This would allow for a significantly lower fee payment by government in place of the lease purchase payment in order to equalize the private sector rate of return at 12 percent.

Outright sale by Government to Private Companies

State and local governments own and operate a variety of facilities that generate revenues in excess of costs—toll bridges, waste disposal facilities, recreation areas, ports, water supply systems, electric and gas utilities, and even transit lines. Sale of these facilities can provide the jurisdiction with substantial revenues. However, private ownership and unregulated pricing may not lead to effi-

Table 28

Return on Investment Under Alternative Service Contract Assumptions Comparison in Each Case: Lease-Purchase at 12% and 14% ROI

Type of Asset		3-Year Asset				5-Year Asset				1-Year Asset			
Depreciation Method		Straight Line		150%		Straight Line		150%		Straight Line		150%	
Cost of Capital[1]		12	14	12	14	12	14	12	14	12	14	12	14
Debt/Equity Ratio	**Residual Value (% of cost)**												
0/100[2]	0	8.5	8.5	9.0	9.0	9.0	9.0	9.5	9.5	7.9	7.9	8.1	8.1
	50	23.6	23.6	24.6	18.0	18.0	18.0	18.7	18.7	10.0	10.0	10.3	10.3
50/50	0	11.2	10.1	12.6	11.5	12.6	11.5	13.9	12.7	10.3	9.1	11.1	9.8
	50	35.9	35.9	38.5	37.8	27.0	26.3	28.7	28.0	13.6	12.8	14.4	13.5
90/10	0	40.4	29.5	77.9	65.2	65.0	54.9	101.0	92.3	62.6	52.2	101.0	70.6
	50	96.2	91.7	125.4	119.8	84.5	79.1	112.1	105.8	63.8	52.8	80.0	70.7

Notes: Costs discounted at the public sector financing rate.

6% ITC assumed on 3-year assets
10% ITC assumed on 5-year assets
No ITC assumed on 15-year assets

1. 12% assumes financing costs equivalent to lease-purchase; i.e., tax-exempt IDB financing available.
   14% assumes private financing available at 2% above lease-purchase rate.

2. Cost of capital not relevant since 100% equity is assumed.

Source: Based on calculations by Jeffrey Apfel, Office of Development Planning, New York State.

cient or equitable outcomes. In some cases—a toll bridge, for example—unregulated private ownership would lead to monopoly pricing which would lead to an inefficiently low level of use, as well as considerable public opposition. In other cases—state-owned office buildings, for example— the problem of monopoly pricing is much less since the facility would have to compete with surrounding privately owned buildings. This section describes the potential financial benefits to be gained by selling a public facility. But the issue of which facilities ought to be sold must be left for each jurisdiction to decide on a case-by-case basis. Clearly, an important element in this decision is the size of the potential revenue enhancement.

The relevant question is, Can the imposition of a tax on a given net revenue stream actually enhance the present value of that revenue stream? An example will help answer the question.

Let us assume the public sector owns a facility that costs $1 million per year to operate but generates annual gross revenues of $2.5 million. The financial value of the asset to the public sector is therefore $1.5 million per year—the amount of the net revenue stream. Let us further assume that the private sector agrees to buy the facility, paying a certain amount in cash up-front and incurring debt for the balance of the purchase price; the public sector borrowing rate is 12 percent; the private sector borrowing rate is 14 percent; and a purchase-money mortgage of $10 million is extended by the public sector at 14 percent over 20 years. The annual debt service to the public sector is $1.5 million—the same amount it would have received had it retained title. But in addition, the public sector will receive an up-front cash payment, the amount of which will be determined by the investor's required rate of return. Assuming the investor pays $2 million in cash on closing, his fifteen-year financial outlook is depicted in table 29. A $2 million cash investment up-front would yield the tax benefits depicted in the last column, resulting in a 12.1 percent rate of return.

The present value of government's net revenues, assuming continued public ownership, (discounted at the 12 percent public sector rate) is approximately $10 million. A year-one cash infusion of $2 million can therefore

Table 29

Cash and Tax Consequences of Sale
46% marginal bracket
(in millions)

	Investment	Income	Expenses	Current Debt-Service Payment	Current Interest	Depreciation	Taxable Income	Tax Benefit	Cash	After-Tax Income
0	−2000								−2000	−2000
1		2500	1000	1500	1400	1200	−1100	506	0	506
2		2500	1000	1500	1386	1080	−966	444	0	444
3		2500	1000	1500	1370	972	−842	387	0	387
4		2500	1000	1500	1351	874	−726	334	0	334
5		2500	1000	1500	1331	787	−618	284	0	284
6		2500	1000	1500	1307	708	−516	237	0	237
7		2500	1000	1500	1280	708	−489	225	0	225
8		2500	1000	1500	1249	708	−458	210	0	211
9		2500	1000	1500	1214	708	−423	194	0	195
10		2500	1000	1500	1174	708	−383	176	0	176
11		2500	1000	1500	1129	708	−337	155	0	155
12		2500	1000	1500	1077	708	−286	131	0	131
13		2500	1000	1500	1018	708	−226	104	0	104
14		2500	1000	1500	950	708	−159	73	0	73
15		2500	1000	1500	873	708	−82	37	0	38

ROI = 12.2%

be considered a 20 percent increase in facility revenues over the fifteen-year period.

This simplified analysis does not, of course, treat several significant issues, including public sector revenues beyond the recovery life, legal mechanisms for sale, and state legal and constitutional issues. But the figures do highlight the potential revenue increment.

State and Local Policy Concerns

In the future, state and local decision makers will be confronted with proposed tax-subsidy finance arrangements. While business tax incentives remain in the federal tax code, and while state and local governments face fiscal problems, these arrangements will remain attractive. This section raises policy concerns that should be considered when evaluating these opportunities. Underlying all the concerns is the assumption that cost is only one factor among many in the public investment decision process.

1. *The importance of public sector control.* Clearly, facilities differ with respect to the need for public control. The public sector may wish to own a facility in order to preserve accountability in management, to protect public employment, or to preserve the external benefits provided by the facility. Some of these concerns can be addressed without maintaining public ownership. For example, the public employment effects of privatization might be lessened if the financing is for a new facility and new employment. Or a private manager might be compelled to provide external benefits through regulation of price or service. It may be helpful to conceive of the instruments discussed in this chapter as lying along a continuum of public control.

Public Control and Financing Instruments

Low need for public control			High need for public control
sale	service contract	lease	safe-harbor lease

At the end of the continuum—low need for public control—is outright sale. While title can be passed to the private sector with strings attached, outright sale normally implies an abdication of basic management responsibilities. Thus, sale can be seen as particularly appropriate where government has assumed quasi-private functions in such areas as real estate and industrial development.

Service contracts, similarly, might be most appropriate only when the public sector is comfortable with the notion of abdicating a relatively high level of management control, since IRS will allow the ITC only when the private sector can document possession and use.

Leases provide a wider degree of public sector control, depending on the terms of the lease. However, current federal restrictions on what constitutes a lease mandate at least some control by the private sector.

Only with the extreme case of safe-harbor leases is private influence negligible, since private ownership is for tax purposes only.

2. *The level of public sector risk.* The public sector may be called upon to assume certain risks in the transactions that must be balanced against the cash generated. One risk is the ability to repurchase. Because non-safe-harbor leases may not contain provisions for nominal public sector repurchase, government must hope that it will have sufficient funds to repurchase a leased facility in the future at the price requested by the lessor. For example, the Suffolk County Sewer District Sale/Leaseback calls for repurchase of the sewer system at market value with reinvested residual bond proceeds. But if today's judgments as to the level of funds required are not accurate, repurchase may be made more expensive or difficult.

Another type of risk the public sector may encounter is indemnification risk. Typically, tax subsidy finance transactions contain provisions to indemnify the investors in the event of legal changes or adverse IRS actions. Such provisions create the possibility that government may be required to compensate the in-

167

vestors in the amount of their lost tax advantages. Because many of the transactions are complex and certain areas of the tax law are untested, this risk may be large.

3. *The effect of the financing on the existing structure of state and local finance.* Leases and service contracts convert what might otherwise be a public operation into a private operation, at least for tax purposes. Thus, the restrictions that many states have on public debt may not apply if the transaction is characterized as private. For example, New York State requires that the principal component of any local debt service payment not exceed by more than 50 percent the principal component of any other debt service payment from the same obligation. This restriction prohibits level-debt service schedules and forces an early amortization of debt. Refinancing a sewer through an IDA, however, allows for level-debt service on the new debt, violating the intent of state law. Additionally, private financing may, like lease purchasing, provide for a way to circumvent constitutional limitations on debt issuance.

4. *The effect on the tax-exempt market.* Changes in institutional behavior, increases in IDB financing, and other factors have already raised tax-exempt rates relative to taxable rates. It is probable that new tax-exempt instruments will exacerbate the problem. In the Suffolk County Sewer District refinancing, $370 million in debt will be retired and $625 million will be issued, all with no new construction to show for it, only a reduction and stabilization of sewer charges at the local level.

5. *Fiscal prudence.* The flexibility afforded by some transactions may allow for a lack of prudence by government or by private firms. Traditional government finance markets impose standards, in part, to assure the long-term soundness of financial systems. Innovation generally brings flexibility, but flexibility can create potential hazards. For example, a municipality, in establishing a fund to reduce local assessments, might be tempted to draw on the fund to subsidize

rates more heavily during the early years of the lease term, leaving much less money to offset assessments in the future. No standards or laws exist to limit this type of borrowing against the future.

6.   *In-state financial incidence.* Not all the tax expenditures at work are borne by out-of-state taxpayers. Depending on the relationship of state and local tax systems to the federal tax system, some portion of the cost may be shifted back to in-state taxpayers.

Conclusion

At a time of increasing needs and decreasing resources, state and local governments should explore the possibilities of tax-subsidy finance. Jurisdictions should carefully examine transactions on a case-by-case basis and, additionally, attempt to set general standards in line with the considerations outlined above. Governments should also recognize that tax-subsidy finance will not be a magic cure for infrastructure problems. It may, however, prove a valuable method, in the short run, for reducing costs.

CHAPTER VII NOTES

1.   This chapter was prepared by Jeffry Apfel, Assistant Director of the Governor's Office of Development Planning, New York State.

2.   The term "facility" in this chapter is used broadly to include not only public structures but other capital items such as transportation equipment, medical equipment and computers.

3.   While "business tax subsidies" will be referred to in this chapter, it should be pointed out that such subsidies may also be available to individuals. Partnerships composed of individuals, for instance, are often the recipients of investment subsidies in the real estate area. Generally, the term "business tax subsidies" will refer, then, to both businesses and individuals.

4.   The system that meets an economist's definition of equity and efficiency is known as the first-year cost recovery system.

5.   A current view is that judicial confusion and disagreement have to

a large extent rendered these standards obsolete, even with respect to equipment. See Freida Wallison, "Benefits of ERTA Available to State and Local Governments," *New York Law Journal,* Vol. 63, No. 1, March 25, 1982.

6. A case for just such an arrangement has been made by Robert Lamb, "The Service Contract vs. The Lease," New York State Department of Transportation, 1982, unpublished.

CHAPTER VIII

Summary and Recommendations

THE NATION IS FACING a crisis precipitated by insufficient and inefficient investment in the development and maintenance of public infrastructure. The crisis can only be averted by radical changes in planning, managing, and financing public capital investments. State and local governments are trapped between the need to arrest the rate of deterioration and to provide for new development and the shrinking resources available to pay for public infrastructure. Failure to increase public investments will jeopardize local and national economic recovery, will weaken governments' ability to deliver public services, and will threaten public safety and health. Yet, the ability of states and localities to raise the necessary revenues is constrained by local revenue and debt limitations, by sharp cuts in federal aid, and by high interest rates.

None of these factors is likely to change significantly in the near future. Budget cutting, small tax increases, and creative financing will not yield the resources to bridge the yawning gap between present levels of investment and what is needed to sustain economic growth through the 1980s.

A comprehensive public capital investment strategy is needed for all levels of government. States can build their strategies around eight elements:

1. Developing cost-effective maintenance, replacement, and design standards for public works projects. The

171

present engineering standards lead to inflated estimates of needs and are insufficient as guides for assigning investment priorities.

2. Reducing public subsidies for private sector investments and the termination of public sector programs and facilities that could be provided by private firms. Scarce public funds must be used to maintain and develop only those projects necessary to fulfill public obligations.

3. Improving planning, budgeting, and management procedures for capital investment. Costly delays, inefficient priorities, and poor management of public facilities inflate costs and accelerate deterioration.

4. Imposing user fees—charges and dedicated tax revenues—to pay for both the construction and operating costs of public facilities. User fees are *not* simply an additional source of revenues and a way to circumvent tax limitations. Efficiently designed user fees are public prices for public goods and services. They will encourage more efficient use of facilities—which will reduce investment needs—and more efficient public investment decisions.

5. Improving bond-financing mechanisms to avoid the proliferation of small issues that increase financing costs. Bond banks, state loans to localities, and state guarantee programs are initiatives that can yield substantial savings.

6. Shifting a larger share of the costs of public facilities associated with large-scale projects onto private developers. For major private projects, state and local governments will have to negotiate the allocation of infrastructure costs with the developers. Also, lease financing and service contracts offer indirect ways of leveraging private sector tax incentives to reduce public sector costs.

7. Negotiating the allocation of capital financing responsibilities among federal, state, and local governments to avoid needless duplication and to ensure that all jurisdictions have adequate resources to make needed investments.

8. Increasing state and local taxes to support increased spending on public works.

None of these steps is easy—either to design or to carry out. Data on the condition of the public capital stock, the cost of repair or replacement, and local priorities are not readily available. There have been few attempts to assess how the changing structure of local economies and rapidly evolving technologies will shape the future demand for public facilities. Yet political leaders can educate the public in the need for new policies only if they can present a coherent and comprehensive public investment strategy.

A durable and effective solution to our deteriorating infrastructure will require initiatives by all levels of government—initiatives undertaken with mutual agreement on the fundamental principles of infrastructure financing. There are four principles that can guide the design of an evolving partnership between federal, state, and local governments.

1. Wherever possible, those who benefit from a particular facility or the services it provides should pay for it. Road users should pay for roads; water users should pay for water; and visitors to a park should pay for the privilege.

2. The capital costs of a facility should be amortized over the expected life of that facility and maintenance costs should not be deferred. This avoids intergenerational transfer of costs.

3. The costs of a facility or program should be paid by the jurisdiction in which the benefits are provided. This reduces the tendency to fund pork-barrel projects. However, some allowance should be made for differences in the fiscal capacities among localities.

4. Where user fees may deny access to a facility or service to the poor, special provisions should be made for low-income households.

These general principles should guide public capital investment policies for all levels of government. In the following sections, the major policy options for state and local

173

governments are briefly discussed. But state and local policies must be developed within the context of federal actions. Many of the factors that have made public infrastructure such an important issue have their roots in federal policies—the poor design of federal grants programs, changes in the tax structure, and mismanagement of the national economy. In the next few years, increased fiscal and economic responsibility for capital investments will devolve onto state governments. This shift is compatible with many of the recommendations discussed in this book. States and localities have the greater potential for obtaining more sound information than federal agencies on what type of capital investments are needed and a much stronger incentive to ensure that investments are wisely made and well managed. But that decentralization must be accompanied by the delegation of necessary powers and the commensurate fiscal resources. For example, the federal government may delegate to states increased responsibility for major transportation programs. But that transfer may be accompanied by other policies to reduce the cost of bond financing and with the delegation of much broader powers to state governments—to impose tolls on sections of the interstate highway or to modify current engineering standards, for example. The National Governors' Association has argued that increased state and local responsibility for "bricks and mortar" programs—projects with relatively local impact—should be balanced by increased federal responsibility for income maintenance and human resource investments, programs that are aimed at national objectives. The final structure of President Reagan's New Federalism initiative is still being negotiated, and the process may take several years to complete. The following section discusses some of the possible changes in federal policies toward infrastructure investment that will shape the way state and local governments develop their own capital investment strategies.

The Federal Context

There are dozens of bills circulating in Congress that would affect the infrastructure issue and the ability of states to finance and manage their public investments.

These initiatives can be divided into four types of federal action.

- [ ] *Subsidies for state and local capital borrowing.* These have been proposed through two mechanisms: (1) low-cost loans from a federal development bank or Reconstruction Finance Corporation or (2) direct interest subsidies of the interest costs of taxable bonds issued by states and localities for public infrastructure projects.

- [ ] *Decentralization of infrastructure financing and administration.* Proposals include greater reliance on block grants, user fees for inland water transportation, and increased revenues into the Interstate Highway Trust Fund, perhaps accompanied by a flexible arrangement with state governments for administration and financing of the system.

- [ ] *A countercyclical public works jobs program.* This could either repeat the Local Public Works Program enacted in 1976 that provided 100 percent federal funding to state and local projects or use a direct federal expenditure program based on depression-era models.

- [ ] *Increased federal planning and managing of public works investments.* Proposals range from a federal capital budget to a reauthorized and expanded Economic Development Administration that would finance state and local planning activities.

Borrowing Subsidies

The purpose of these initiatives is to address the weakening of the tax-exempt bond market, which resulted from tax reforms and changing behavior by institutional investors, and the deteriorating fiscal positions of state and local governments. The two types of recommendations are (1) to provide low-interest-rate loans to state and local governments for major capital investment projects and (2) to subsidize directly the interest costs on taxable infrastructure bonds issued by state and local governments.

Federal Loans. A reauthorized EDA or a Reconstruction Finance Corporation have been proposed to make low-interest-rate loans and/or capital grants to municipalities and states for major capital projects. Under most proposals, the subsidies would be at the discretion of the administering agency and would be negotiated. The federal loans would be financed through the issue of federal debt (which may exacerbate the competition with municipal borrowing in the capital market) and by direct appropriations to provide grants and interest write-downs. It would be, in essence, a very-large-scale Urban Development Action Grant program.

The proposal would allow federal agencies to ensure that local capital spending was targeted at areas of national need and would extend the fiscal capacity of local jursidictions. But the initiative may be subject to the same problems that have surrounded many similar, if smaller, federal programs in the past. First, it may encourage states and localities to develop projects that would receive federal aid rather than projects that are needed and to "over-design" the projects in order to prove that they could not be paid for out of state and local resources. Second, year-to-year funding would create considerable uncertainty. Third, ensuring a regional allocation of funding would lead to different needs criteria in different areas. Fourth, some projects may be funded for political rather than economic reasons. Finally, the program would be very expensive.

Taxable Bond Option. During the last decade, there were several proposals, both by Congressional committees and by analysts of the municipal bond market, to provide state and local governments with the option of issuing taxable bonds, for which they would receive a direct subsidy equal to 30 or 35 percent of the interest costs from the U.S. Treasury. The proposal was argued for three reasons.

☐ First, there is growing belief that exemption of interest from personal income taxes was unnecessarily expensive. Some research indicates that as much as one-third of the revenues not collected by the U.S. Treasury are essentially subsidies to high-income individuals (Vaughan, 1980). A direct subsidy from the U.S. Treasury could be designed that would be cheaper in terms of revenues lost and would provide a greater borrowing subsidy to state and municipal governments.

☐ Second, a taxable bond would compete in a much broader capital market than a tax-exempt bond and would, therefore, make it easier to market public debt.

☐ Third, the taxable bond market is more stable than the tax-exempt market; therefore, a public taxable bond would offer the public sector the opportunity for more stable long-term financing.

However, several studies have called into question these arguments. Mussa and Kormendi (1979) argue that those that have asserted that the exemption benefits very-high-income households at the expense of low-income households are wrong because at least two-thirds of all tax-exempt bonds are held by institutional investors (pension funds, trust funds, insurance companies, and so forth) whose investments benefit a very broad class of income groups. Mussa and Kormendi also question the efficiency of providing a deeper subsidy for state and municipal borrowing. However, their arguments were made before the Economic Recovery Tax Act of 1981, before the withdrawal of institutional investors from the tax-exempt market, and before the sharp reductions in federal grants for public infrastructure. For these reasons, the proposal is being reconsidered.

The proposal is fairly simple and inexpensive. States and localities would be empowered to issue taxable bonds and would receive an annual cash payment of between 25 and 35 percent of the interest cost from the U.S. Treasury. Since the marginal tax rate of bond purchasers is at least 35 percent, the tax revenues collected by IRS would be about equal to the cost of the subsidy (Koepke and Kimball, 1979). If the long-term taxable bond rate were 13 percent, the post-subsidy cost to a locality would be 8.45 percent, nearly 200 basis points below the 10.4 percent on tax-exempt bonds that has prevailed during the last quarter of 1982.

Three features could distinguish most planned versions of this from past proposals. First, the taxable bond would be an option for state and local governments. It would not replace tax-exempt bonds. Second, the taxable bond option would be made available for only certain types of public investments—including public education and health facilities, roads, water treatment facilities, and ports and terminals. It could explicitly exclude public bonds whose revenues were used to finance private facilities, or facilities with private tenants—including not-for-profit health care institutions, industrial facilities, and pollution control projects for private firms.

Third, the interest subsidies would be an entitlement and not be subject to annual budget appropriations. Placing the subsidy as a line budget item would make it vulnerable to the vagaries of the annual appropriations process, which will create sufficient uncertainty to deter rational long-term planning by state and local governments. The taxable bond option may be offered as an entitlement to states and localities in exchange for increased state assumption of financing responsibilities.

The cost of the program would be borne by high-income households whose 50 percent marginal tax rates make municipal bonds at 10 percent the equivalent of a taxable bond earning 20 percent. The program is less cumbersome than categorical grant programs or a direct federal loan initiative because it would entail no expensive allocation procedure or costly mandates and regulations.

Decentralization of Infrastructure Responsibilities

Decentralization of both fiscal and administrative responsibilities for infrastructure construction and maintenance seems inevitable. State governments will have to pay a larger share of the costs and undertake more of the planning, management, and administration of the facilities and systems. From the state perspective the important issues are (1) what fiscal responsibilities are assumed by the federal government, (2) what programs are decentralized, and (3) what funding sources accompany the decentralized programs.

The first issue is too broad to be discussed here. The second question is also highly complex because more than

400 federal categorical programs remain on the books, many of which have a capital facilities component. Two of the most important broad federal programs with the greatest impact on infrastructure are water projects and the interstate highway system (including the repair of bridges).

Water projects. Most major water projects concerned with inland freight transportation are 100 percent financed and managed by the Army Corps of Engineers. Many water projects concerned with electricity generation, irrigation, and municipal water supply are undertaken by the Bureau of Reclamation. Studies by the U.S. General Accounting Office have found that some of these projects are not cost-effective. Because no state or local funds are involved, localities have no incentive to identify and undertake only those projects that are economically viable. Senators Moynihan and Domenici have proposed to convert water project funding into a block grant program, which would require a state match for each project funded. This proposal is consistent with the administration's New Federalism and, since it could yield substantial savings to the federal government, may receive more attention in the future. State administration of these funds could be effectively undertaken through a statewide water authority that would issue bonds, backed by user fees, to undertake major projects and to lend money to localities. New York State has already completed feasibility studies and proposed enabling legislation for such an authority. The experience of irrigation districts in many western states also provides useful models.

Decentralization of the Interstate Highway Program. The deteriorating condition of the interstate highway system will be one of the most expensive infrastructure programs during the next decade. The four cents per gallon gasoline tax, unchanged in two decades, is no longer adequate. Inflation in construction costs and the switch to smaller cars has eroded revenues drastically. Replacement, reconstruction, and rehabilitation costs are escalating rapidly. The increase in the tax, enacted in December 1982, was a necessary element in developing an effective solution (CBO, 1982). A long-term solution might also include a decentralization of management and administration of the system that would allow states to set up their own trust funds that would receive federal monies but would be augmented by revenues from state gas taxes, tolls (if the present requirement that tolls be phased out is relaxed), and other dedicated fees and taxes.

Allocations to the state trust funds could be made in a manner similar to the present allocations, with a state match required (which could include toll revenues), but the states would have much greater flexibility over choosing which, among presently eligible projects, to undertake. States that elected this option could choose which incomplete sections to complete, which bridges to repair, and which sections to upgrade, ending the time-consuming and highly political negotiations that currently attend these decisions. All states would be required to set up trust funds by some prescribed date, and all federal highway expenditures would be channeled through these funds. The trust fund structure would allow long-term comprehensive planning at the state level. Washington could define minimum standards to ensure the maintenance of a uniform system. The combination of increased gasoline tax revenues, tolls, and enforced local setting of project priorities could provide for a long-term solution to crumbling roads.

Countercyclical Public Works

The high rate of unemployment—it approached one quarter of all employees in the construction industry in September 1982—has led to a proliferation of legislative proposals to provide federal funds for public works projects. Such a program might be modeled on the Local Public Works Program (LPWP), initiated in 1976, which provided 100 percent federal financing through grants to state and local governments, or modeled on President Roosevelt's Works Progress Administration (WPA), under which the federal government directly hired workers for federally financed projects.

Past experience suggest that neither approach would provide a long-term solution to the infrastructure problem, although they may provide some fiscal relief to state and local government (see Vernez and Vaughan, 1978). The LPWP was beset with many problems (summarized in Vaughan, 1980). Most of the funds were not spent until four years after the 1975 recession, and the average job created lasted only a month. There was little increase in state and local capital spending, since federal funds simply supplanted local spending. In fact, there was a sharp decline in state and local spending, while these governments waited to see which of their project applications would be federally financed. And many of the projects undertaken were paid for because they could be started quickly, not because they were a high local priority. The WPA model may be very difficult to undertake today because public construction requires a labor force versed in skills that few of the unemployed have acquired.

Increased Federal Planning and Management

The mounting concern over the deterioration of public works and the lack of data from which to assess the extent of the problem and to evaluate alternative solutions has led to several proposals to increase the public capital planning and managing activities of the federal government. These include the development of a national capital budget that would divide federal expenditures between operating and capital and would calculate the rate of depreciation of existing capital facilities, federal planning grants to states and localities, and investment strategies negotiated between federal agencies and state and local governments.

The federal government may help develop new techniques for setting user fees and improve other analytic techniques discussed in academic journals but not applied to state and local decision making. The privatization of some public activities will only be possible after some complex research, reform of regulatory procedures, and the establishment of markets for public goods and services where none currently exist. This, too, could be aided by a federal commitment to funding research and demonstration projects.

The dissemination of information may be a function

best performed by a federal agency. State and local governments should be able to draw upon data on technological innovations in construction and maintenance practices, on needs assessment techniques, and on evaluations of policy and program innovations carried out by other states and areas.

The State Role

The greatest increase in fiscal and administrative responsibility for public capital investments will fall upon state governments. Not only will states have to change their own practices and levels of effort, but they will have to provide increased assistance—fiscal, technical, regulatory, and administrative—to local units of government. Many of the state functions will be unfamiliar—regulating private water suppliers, entering into leasing arrangements with private firms, or managing hazardous waste disposal sites, for example. All actions will require public education and leadership to secure the passage of enabling legislation or budgetary obligations against the efforts of individual interest groups who have benefited from implicit and explicit subsidies in the past and who will face higher taxes or user fees in the future. It will be necessary to present a comprehensive strategy in which the general principles and all the elements of them are outlined.

Setting the State House In Order

The first step in developing an effective state capital investment strategy is to develop state capital budgets. Increased spending must be based upon much more careful analysis of needs and costs of alternative actions. Capital budgets should include more than a wish list of future projects. They must include a discussion of the underlying values and principles that guide the strategy, an analysis of public and private sector responsibilities, a discussion of the mechanisms that will be used to finance both construction and operating costs, and a careful analysis of the impacts of capital projects on future operating expenditures. Without this long-term framework (discussed in detail in volume 1 of this series, *Planning and Managing*),

Table 30

User Fees for State Infrastructure Programs

Type of Infrastructure	Possible User Fees
Road Systems	— Ad valorem vehicle registration fee — Gasoline taxes — Other transportation-related sales taxes — Bridge and highway tolls (peak load pricing)
Research and Higher Education	— Tuition assistance with "royalty payback" — State match to private R&D grants — Patent policy reform — Tax incentives for corporate donations — Technology user fees
Transit	— Increased fares (especially during peak hours) — Subsidy from gasoline tax — Special district assessments — Para-transit licensing
Water Supply	— Metered water use — Full-cost pricing for utility hook-ups — Peak-load pricing
Waste Water Treatment	— Industrial and commercial effluent charges
Waste Disposal	— Full-cost coverage from disposal fees
Ports and Terminals	— Full-cost coverage from fees and charges to users
Recreation	— Peak-load pricing of seasonal facilities — Special assessment district revenues — Tax increment financing

a sound strategy will not be designed nor will consistent principles be followed.

The state must undertake a long-term review of the way it manages its public facilities, including assessing whether the institutional framework (state agency, public authority, special district, and so forth) leads to efficient cost control, what type of analytic techniques are used to establish investment needs, and what revenue sources are available. Public referendums will be required for some actions (tax increases, tax dedication, bonding power for public authorities, and so forth), a process that takes several years.

States can undertake specific policies to address infrastructure needs. These actions will include:

☐ Privatization of existing public facilities and services (accompanied by appropriate changes in regulatory policy).

☐ Reduced use of public bond issue revenues to subsidize private projects.

☐ Closing discretionary tax abatement and exemption programs and using the increased revenues to finance infrastructure programs.

☐ Imposition of user fees for a broad array of state capital investment programs (see table 30).

☐ Increased use of private tax incentives to finance public infrastructure.

☐ Negotiated investment strategies with developers of large-scale projects, especially in small communities.

These changes may involve the loss of discretionary power over the annual budget process and will challenge the states to develop mechanisms to make public authorities and the increasing numbers of private suppliers of facilities and services fully accountable to state agencies. But without these changes, decisions with long-term implications will continue to be made with short-term procedures. And capital projects will be undertaken for reasons of political expediency rather than as sound economic investments.

State Assistance to Localities

In addition to reforming their own planning, budgeting, and financing techniques, state governments will have to work much more closely with localities to develop a coordinated investment strategy. The state can ensure that local financing and management policies do not lead to interjurisdictional pricing competition and that local investment practices are consistent with broad state objectives. The state must increase its expenditures on research and evaluation of public capital investment needs and of alternative financing mechanisms, and share the results with localities. Many of the issues that surround the design of an effective user fee system, the transfer of responsibility to the private sector, or the negotiation of cost-allocation with private developers are complex and are beyond the capacity of local government units.

Some states will also have to amend the taxing powers of local governments to allow appropriate user fee systems and create more flexible local administrative mechanisms to make public investments. But this delegation of responsibility must be accompanied by increased state oversight of local government policies and practices. Procedures must be developed to allow more flexible negotiations between state and local governments to establish investment priorities.

The sharp reduction in federal aid will create special problems for jurisdictions with high concentrations of low-income households and with distressed local economies. States must review their intra-state revenue-sharing programs and sharpen the targeting where possible. Categorical aid programs could also be refocused, with the state share to nondistressed communities cut back and aid to needy areas increased.

State regulatory policy should also be carefully reviewed for its effects on the costs of local public works investments. State labor law, contract-letting procedures, tax policies, permitting procedures, and construction codes can all contribute to inflated costs and unnecessary delays (NCSL, 1982).

Table 31 lists the major initiatives that states can undertake to assist localities in reducing infrastructure financing costs. While these measures will not dramatically reduce debt issuing and servicing costs, programs such as bond banks and low-interest-rate loans can prove very valuable to distressed communities. They can also help create an atmosphere of cooperation between states and their component jurisdictions that will allow for more radical, and sometimes painful, steps to be taken.

185

Table 31

Possible Actions for State Governments*

Action	Suitable States
Improve technical assistance programs.	All
Create a loan program for water & sewer construction.	All
Create a municipal bond bank.	Rural states with many small issuers
Earmark state aid for debt service.	States with fiscally distressed communities
Assist local governments with creative financing (through technical assistance Programs and enabling legislation).	All
Create loan programs for energy impact assistance.	Energy-rich states
Increase state supervision of local debt management.	States with cities having poor credit ratings or histories of poor financial management.
Guarantee local debt.	Most states

Note: *The most effective measures are listed first.

Source: National Conference of State Legislatures, Watson, 1982.

Benefits	Drawbacks
Facilitate bond issuance and encourage responsible debt management.	Local governments may fear state intrusion.
Stimulate investment in water and sewer facilities, and supplement federal wastewater program. Provide loans at favorable rates, particularly for distressed communities and small or infrequent Issuers.	Increase state general obligation debt and thereby lower state credit rating.
Reduce borrowing costs for small or infrequent Issuers.	Local banks, bond counselors, and underwriters may suffer a loss in business.
Improve rating of bond issues (reduce interest rate) for large, fiscally distressed communities.	Requires a large, permanent state aid program and some state supervision.
Increase local flexibility. Facilitate use of beneficial techniques, and discourage improper use.	Many techniques are untested. May result in excessive short-term debt, or high interest rates in future.
Finances rapid capital construction in energy boom towns. Supplement to grant programs.	Local governments prefer grants or sharing of severance revenues.
Encourage responsible debt management, and improve credit ratings.	Increased administrative costs for local governments, possible restriction on local actions.
Improve rating on local bonds.	May seriously weaken state credit rating.

Conclusion

There is no simple and cheap way to pay for public works. There is no easy way to cut back on existing services and programs, to charge user fees, or to raise taxes. Yet failure to devise a rational and effective way of investing in public works will surely prevent any significant improvement in the national and state economies. We must use this crisis as an opportunity to redefine public investment priorities. A sound fiscal strategy and a clear allocation of responsibility between the public and private sectors are much more powerful development incentives than speculative projects, tax subsidies, and uncertainty over the condition of vital public works. Sustained economic growth will require increases in taxes to pay for the maintenance and development of roads, bridges, ports, water supply and treatment systems, and facilities for the disposal of hazardous waste. States can no longer afford to delay designing a complete capital investment strategy.

Annotated Bibliography

The following annotated bibliography is intended as a brief guide to the overall bibliography. It classifies references under the major topics covered in this book.

For overall statistical data on public investments the most useful sources are the *Survey of Government Finances,* annually; various articles in the *Survey of Current Business* (monthly, published by the Bureau of Economic Analysis); the Office of Management and Budget; and various reports by the Bureau of Economic Analysis. There are also reports published by various interest groups including the American Public Works Association, Federal Highway Users Association, American Public Transit Association, and Association of Metropolitan Sewage Agencies.

The best single source on public capital finance issues is the Municipal Finance Officers' Association, which has an extensive set of publications. The National Conference of State Legislatures, the Council of State Planning Agencies, and the Urban Institute also have extensive sets of publications and reports.

Bond Banks:

Forbes and Renshaw, 1972; Jarrett and Hicks, 1977; Katzman, June 1980 and March 1980; Solano and Hoffman, 1982.

Bonds: *(see also Bond Banks, Credit Ratings, Debt, Housing Bonds, Industrial Development Bonds, Taxable Bonds)*

For an overview of municipal bonds and definitions and descriptions of the bond market see ACIR, 1976; Amdursky, 1981; Lamb and Rappaport, 1980; Peterson, J., 1976; Public Securities Association, 1981. See also Durst, 1981; Fischer et al., 1980; Klapper, 1980; Robinson, 1981; and White, 1979. Basic data on bonds and bond markets is reported in the *Daily Bond Buyer* and in *Moody's Bond Survey* (weekly) and is summarized in the annual reports of the Public Securities Association.

For an analysis of the behavior of the bond market, including the effects of federal tax changes, see Browne and Syron, 1979; Forbes and Peterson, 1975; Hendershott and Koch, 1977; Herships and Karvelis, 1981; Kaufman, 1981; Kimball, 1977a; Kopcke and Kimball, 1979; Mumy, 1978; Peterson, J., 1982; Twentieth Century Fund, 1976; Viscount, 1982; and U.S. Congress, Joint Economic Committee, 1981.

Insurance: Miralia, 1980.

Marketing: MFOA, 1976.

New Jersey Qualified Bond Program: Jones, 1978; Peterson and Miller, 1981.

Small Denominations: Lehan, 1980.

Underwriting: Cagan, 1978; Silber, 1980.

Boom Towns: *(see also Energy Development)*
Brookshire and D'Arge, 1980; Gilmore and Duff, 1975; O'Hare, 1977; O'Hare and Sanderson, 1978; U.S. Department of Defense, Office of Economic Adjustment, 1981a and 1981b.

Capital Budgeting: *(see also Debt, Management and Planning)*
For a discussion of the process and benefits of capital budgeting, see Choate and Walter, 1981; Devoy and Wise, 1979; Douglas, 1977; Fujardo, 1976; Howard, 1973; Matson, 1976; U.S. GAO, 1982, 1981, and 1980; Wacht, 1980; and White, 1978. For examples of state initiatives, see Alexander, 1980; American Public Works Association, 1979; Maryland, 1980; and Rumowicz, 1980.

Condition of Public Works: *(see also Financing Public Investment)*
For a brief account of the problem, see Choate and Walter, 1981, and *Newsweek* 8/2/82. For more detailed discussions of the extent of the problem, see Abt Associates, 1980; CONSAD, 1980; Dossani and Steger, 1980; U.S. Department of Commerce, 1980. The study of the capital stock in 12 cities conducted by the Urban Institute is exhaustive: see Grossman, 1979; Humphrey and Wilson, 1980; Humphrey et al., 1979; Peterson and Miller, 1981; Wilson, 1980. See also American Public Works Association, 1981 and 1976; Beals, 1981; Buckley, 1982; Choate, 1982; Finck and Pike, 1981; Hatry, 1980; Lindsay, 1979; and MacDonald, ed., 1982.

Countercyclical Public Works:
See CBO, 1978; Kaus, 1982; Vaughan, 1980a, 1980b, and 1976; and Vernez and Vaughan, 1978.

Credit Rating: *(see also Bonds, Debt, Fiscal Capacity)*
The most comprehensive guides are Ingram and Copeland, 1982; Peterson, J., 1974; Smith, 1979. See also American

Bankers Association, 1968; Aronson and Marsden, 1980; Boyett and Giroux, 1978; Osteryoung, 1978; Reilly, 1967.

Debt: *(see also Bonds, Credit Ratings, Fiscal Capacity)*
ACIR, 1976; Gold, 1981; Irwin, 1979.
Danger Signals: Aronson, 1976.
Limitations: ACIR, 1980; Baer, 1981.
Management: Brown et al., 1978; Moak, 1970; Peterson, J., 1979a, 1978; Small Cities Financial Management Project, 1978; Steiss, 1975. State Assistance to Localities: Alaska Legislative Affairs Agency, 1967; Forbes and Peterson, 1978; Glaser, 1978; Peterson, 1977 (see also Bond Banks).

Energy Development: *(see also Boom Towns)*
Foster, 1977; Kolb, 1982; Leistritz and Murdock, 1981; Lu, 1977; Monaco, 1977; Rocha, 1982; Sanderson, 1977; Schnell and Krannich, 1977; Susskind and O'Hare, 1977; and West, 1977.

Federal Government: *(see also Bonds, Intergovernmental Relations)*
Grants: Executive Office of the President, 1980 and 1978; National Governors' Association, 1977 (see also Intergovernmental Relations).
And Local Fiscal Conditions: Herships and Karvelis, 1981 (see also Fiscal Capacity).
Impact on Infrastructure: Gramlich and Galper, 1973; Hatry, 1980.

Financing Public Investment: *(see also Capital Budgeting, Condition of Public Works, Lease Financing, User Fees)*
For summaries of policy alternatives, see Pagano and Moore, 1980; Peterson, n.d.; Peterson and Miller, 1981; Stanfield, 1980; Watson, 1982. For bibliography, see Buss, 1981. See also Buckley, 1982; Getzels and Thurow, 1980; McWatters, 1979; Saffran, 1979; Shaul, 1981; Wolman and Reigeluth, 1980.

Fiscal Capacity: *(see also Bonds, Debt, Intergovernmental Relations)*
ACIR, 1971; Crinder, 1978; Herships and Karvelis, 1981; Howell and Stamm, 1979; Hubbell, 1979; Matz, 1980; Rothschild, Unterberg, Towbin, 1982; U.S. Joint Economic Committee, 1981; and Wolman and Davis, 1981.

Highways: *(see also User Fees)*
For a discussion of financing, see Congressional Budget Office, 1982; Langton, 1981; "Forty States Eye Motor Fuel Tax Boost," 1981.

Housing Bonds: *(see also Bonds)*
Bates and Wolfson, 1981; Harrington, 1979; Levatino-Donoghue, 1979; National Conference of State Legislatures, 1980; Peterson, G., 1979; Worsham, 1980.

Industrial Development Bonds:
Edmonds and Hoyd, 1981.

Intergovernmental Relations:
For an overview, see ACIR, 1980, 1979, and 1978. See also CBO, 1978, and Zimmerman, 1976.

Lease Financing: *(see also Financing Public Investment)*
Barnes, 1981; Dyl and Joehnk, 1978; Lubick and Galper, 1982; Mentz et al., 1980; Schellenbach and Weber, 1978; Shubnell and Cobb, 1982.

Management and Planning: *(see also Capital Budgeting)*
Amara, 1979; American Society of Planning Officials, 1977; Bologna, 1980; Corr, 1980; Donnelly, 1980; Hall, 1979; Hatry, 1980; Higgins, 1978; Holloway and King, 1977; International City Managers Association, 1981; Korbitz, 1976; Kunde and Berry, 1982; Naylor, 1979; Naylor and Neva, 1980 and 1979; Schmidt, 1979; Schneider and Swinton, 1979; Sheeran, 1976; Shepard and Goddard, n.d.; and Steiss, 1975.

Public Authorities:
Beyer, 1972; Edelman, 1976; Holland, 1972; and, for the most detailed analysis, Walsh, 1976.

Special Assessment:
Shoup, 1980.

Taxable Bond Option:
Kimball, 1978; Kopcke and Kimball, 1979; Mussa and Kormendi, 1979.

Taxes:
And Economic Development: Kieschnick, 1981, and Vaughan, 1979.
Of Mineral Resources: Zeller, 1982.
Limitations on Revenues: Bacon, 1981; Baer, 1981; Benson, 1980; McWatters, 1979; Pascal, 1980; Saffran, 1979.

User Fees:
The best overview of the principles and applications of user fees is in Mushkin, 1972. See also ACIR, 1974; Buss, 1981; Downing, 1980; Feldstein, 1972; Galambos and Schreiber, 1978; Gold, 1979; Mick, 1981; Mushkin, 1979 and 1977; Mushkin and Vehorn, 1977; Stanfield, 1980.

Airports: Feldman, 1967, Littlefield and Thompsen, 1977.
Health: Badgley and Smith, 1979, Berki in Mushkin, ed., 1977.
Highways: Abouchar, 1974; Feuer, 1978; Henion and Ford, 1981; Higgins, 1979; Langton, 1981; Smith, 1980.
Recreation: Artz and Bermond, 1970; Economics Research Associates, 1979.
Residential Infrastructure: Bacon, 1981.
Transit: Institute for Public Administration, 1980.
Waste Disposal: Albert, Hansen, and Wilkinson, 1972; Dales, 1970; DeLucia, 1974; Hanke and Wentworth, 1981.
Water: Angelides and Bardach, 1978; Barry, 1976; Carey, 1976; Grover, 1980; Hanke, 1981 and 1976; Hoggan, 1977; Keller, 1977.

Water: *(see User Fees)*
Grossman, 1979b; Keller, 1977; Kish, 1980; Lake, 1979.

Bibliography

Abt Associates, Inc., *National Rural Community Facilities Assessment Study*, Boston, MA, 1980.

Abouchar, Alan, "A New Approach to the Evaluation and Construction of Highway User Charges," *Eastern Economic Journal*, 1974, pp. 34–38.

Adams, Charles F., Jr., and Dan L. Crippen, "The Fiscal Impact of General Revenue Sharing on Local Governments," unpublished report prepared for the Office of Revenue Sharing, U.S. Department of the Treasury, November, 1978.

Advisory Commission on Intergovernmental Relations, *Significant Features of Fiscal Federalism, 1979–80 Edition*, Washington, DC: GPO, 1980.

_____, *Restructuring Federal Assistance: The Consolidation Approach*, Bulletin No. 79–6, October, 1979.

_____, *The Intergovernmental Grant System: An Assessment and Proposed Policies*, (B-1), September, 1978.

_____, *State Mandating of Local Expenditures*, (A-67), July, 1978.

_____, *A Catalog of Federal Grant-in-Aid Programs to State and Local Governments: Grants Funded FY 1975*, (A-52a), October, 1977.

_____, *Categorical Grants: Their Role and Design*, (A-52), May, 1977.

_____, *Improving Federal Grants Management*, (A-53), February, 1977.

_____, *The Intergovernmental Grant System as Seen by Local, State and Federal Officials*, (A-54), March, 1977.

_____, *Community Development: The Workings of a Federal-Local Block Grant*, (A-57), March, 1977.

_____, *The States and Intergovernmental Aids*, (A-59), February, 1977.

————, *Federal Grants: Their Effects on State-Local Expenditures, Employment Levels, and Wage Rates*, (A-61), February, 1977.

————, *Pragmatic Federalism: The Reassignment of Functional Responsibility*, (A-49), July, 1976.

————, *Local Revenue Diversification: Income Sales Taxes and User Charges*, (A-47), June, 1974.

————, *Multistate Regionalism*, (A-39), April, 1972.

————, *Measuring the Fiscal Capacity and Effort of State and Local Areas*, (M-58), March, 1971.

Alaska, Legislative Affairs Agency, *State Assistance to Local Governments on Bonding Problems*, Juneau, AL: Legislative Council, Legislative Affairs Agency, January, 1967.

————, Legislative Budget and Audit Committee, *Alaska's Public Corporations: A Framework for Assessment*, prepared by the Institute for Public Administration, New York, January, 1982.

Aldrich, Mark, *A History of Public Works in the United States, 1790–1970*, Washington, DC: U.S. Department of Commerce, 1979.

Alexander, Governor Lamar, *The Five Year Capital Budget for the State of Tennessee. 1980–81, 1984–85*, Nashville: The Government of the State of Tennessee, January, 1980.

Amara, Ray, "Strategic Planning in a Changing Corporate Environment," *Long-Range Planning*, Vol. 12, February, 1979, pp. 2–16.

Amdursky, Robert S., *Municipal Bond Law: Basics and Recent Developments: A Course Handbook*, New York: New York Practicing Law Institute, 1981.

American Bankers Association, Bank Management Committee, *A Guide for Developing Municipal Bond Credit Files*, New York, 1968.

American Enterprise Institute, *Waterway User Charges*, Washington, DC, 1977.

American Public Works Association, *Revenue Shortfall: The Public Works Challenge of the 1980s*, Chicago, 1981.

————, *Administration of State Capital Improvement Programs: Nine Selected Profiles*, Chicago, 1979.

_____, *A History of Public Works in the United States*, Chicago, 1976.

American Society of Planning Officials, *Local Capital Improvements and Development Management: Literature Synthesis*, Washington, DC: GPO, July, 1977, prepared for the U.S. Department of Housing and Urban Development and the National Science Foundation.

Angelides, Sotirios, and Eugene Bardach, *Water Banking: How to Stop Wasting Agricultural Water*, San Francisco: Institute for Contemporary Studies, 1978.

Aronson, J. Richard, *Determining Debt's Danger Signals*, International City Management Association, Management Information Service, Vol. 8, no. 2, December, 1976.

Aronson, J. Richard, and James R. Marsden, "Duplicating Moody's Municipal Credit Rating," *Public Finance Quarterly*, Vol. 8, no. 1, January, 1980, pp. 97–106.

Aronson, J. Richard, and Eli Schwartz, *Management Policies in Local Government Finance*, Washington, DC: International City Managers Association, 1975.

Artz, Robert M., and Hubert Bermond, "The Fee" in *Guide to New Approaches to Financing Parks and Recreation*, New York: Acropolis, 1970.

Association of the Bar of the City of New York, Committee on Municipal Affairs, *Local Finance Project: Proposals to Strengthen Local Finance Law in New York State*, New York, November, 1978.

Bacon, Kevin, "Paying for Public Facilities after Proposition Thirteen," *Western City*, August, 1981, pp. 8–11.

Badgley, Robin F., and R. David Smith, *User Charges for Health Services*, Toronto: Ontario Council of Health, 1979.

Baer, Jon A., "Municipal Debt and Tax Limits: Constraint on Home Rule," *National Civic Review*, Vol. 70, no. 4, April, 1980, pp. 204–10.

Barnes, Garry, "Going Into Lease Financing," *The Bankers Magazine*, Vol. 164, no. 4, July/August, 1981, pp. 9–14.

Barr, James L., "Rational Water Pricing in the Tucson Basin," *Arizona Review*, Vol. 25, October, 1976.

Bates, John C., and Barry M. Wolfson, "New Mortgage Bond Arbitrage Restrictions: The Problems and Potential Solutions," *Weekly Bond Buyer,* September, 1981, pp. 5, 45.

Beals, Alan, *Cities: Infrastructure Problems and Needs,* Washington, DC: National League of Cities, September 10, 1981.

Benson, Earl D., "Municipal Bond Interest Cost, Issue, Purpose and Proposition 13," *Governmental Finance,* September, 1980, pp. 15–19.

Beyer, Stuart, *Statewide Public Authorities in New York: The Question of Control,* Ph.D. Thesis, State University of New York at Albany, University Microfilms, NY, 1972. (Ann Arbor, MI).

Bologna, Jack, "Why Managers Resist Planning," *Managerial Planning,* January/February 1980, pp. 51–56.

Boyett, Arthur S., and Gary A. Giroux, "The Relevance of Municipal Financial Reporting to Municipal Security Decisions," *Governmental Finance,* May, 1978, pp. 29–34.

Braun, J. Peter, et al., "Dollars for Debt: A Case for Planned Debt Management," *New Jersey Municipalities,* November, 1978, pp. 1–7.

Brody, Susan E., *Federal Aid to Energy Impacted Communities: A Review of Related Programs and Legislative Proposals,* Cambridge, MA: Lab of Architecture and Planning, MIT, 1977.

Brookshire, David S., and Ralph C. D'Arge, "Adjustment Issues of Impacted Communities or, Are Boomtowns Bad?" *Natural Resources Journal,* Vol. 20, July, 1980.

Browne, Lynn E., and Richard F. Syron, "The Municipal Market Since the New York City Crisis," *New England Economic Review,* July, 1979, pp. 11–26.

Buckley, Michael Patrick, "Assessing the Issues and Trends in Public Utilities Financing: Planning and Policy Considerations for State and Local Governments in Oregon," master's thesis, University of Oregon, Salem, 1982.

Burchell, Robert W., and David Listokin, *Cities Under Stress,* New Brunswick, NJ: Rutgers University, Center for Policy Research, 1981.

Burr, Eugene, *Preparation of a Capital Improvement Program,* MTAS Technical Report, Knoxville: Municipal Technical Advisory Service, Institute for Public Service, University of Tennessee Municipal League, November, 1975.

Buss, Terry F., *Innovative Financial Mechanisms for Urban Economic Development: A Bibliography*, Public Administration Series, Monticello, IL: Vance Bibliographies, May, 1981. Bibliography P-732, p. 14.

_____, *Public/Private Partnerships for Urban Economic Development: A Bibliography*, Public Administration Series, Monticello, IL: Vance Bibliographies, May, 1981. Bibliography P-731.

Cagan, Phillip, "The Interest Savings to States and Municipalities from Bank Eligibility to Underwrite All Nonindustrial Municipal Bonds," *Government Finance*, May, 1978, pp. 40–48.

Carey, D.I., "Conservation Water Pricing for Increased Water Supply Benefits," *Water Resources Bulletin*, Vol. 12, December, 1976, pp. 111–123.

Centaur Associates, Inc., *Economic Development Administration Title 1 Public Works Programs Evaluation*, Washington, DC, May, 1979.

Choate, Pat, "Special Report on U.S. Economic Infrastructure," unpublished paper, The House Wednesday Group, May 18, 1982.

_____, *As Time Goes By: The Costs and Consequences of Delay*, Columbus, OH: The Academy of Contemporary Problems, 1980.

_____, "Urban Revitalization and Industrial Policy: The Next Steps," testimony before the Subcommittee on the City, Committee on Banking, Finance, and Urban Affairs, U.S. House of Representatives, hearings on Urban Revitalization and Industrial Policy, Washington, DC, September 7, 1980.

Choate, Pat, and Susan Walter, *America in Ruins: Beyond the Public Works Pork Barrel*, Washington, DC: Council of State Planning Agencies, 1981.

Citizen's Budget Commission, Inc., *A Plan to Expedite the Rebuilding of New York City's Infrastructure*, New York, 1979.

Clyde, Larry F., "Statement before the Senate Committee on Finance," U.S. Senate, Washington, DC, 1982.

Coltman, Edward, and Shelley Metzenbaum, *Investing in Ourselves: Strategies for Massachusetts: A Report to the Task Force on Public Pension Investments of the Massachusetts Social and Economic Opportunity Council*, Conference on Alternative State and Local Policies, Washington, DC, 1979.

Congressional Budget Office, *The Interstate Highway System: Issues and Options,* Washington, DC: GPO, June, 1982.

———, *Countercyclical Uses of Federal Grant Programs,* Washington, DC: GPO, November, 1978.

Congressional Research Service of the Library of Congress, *Review of Title V Commission Plans,* Washington, DC: GPO, 1977.

CONSAD Research Corporation, *A Study of Public Works Investment in the United States,* Washington, DC: GPO, 1980. In volumes.
- Vol. 1 *Historical Analysis of PWI Trends and Financing Mechanisms.*
- Vol. 2 *Analysis of Maintenance, Condition and Financing of Urban Capital Stock.*
- Vol. 3 *Effects of Federal Capital Grants on the State-Local Functions: Water Systems, Sewer Systems, Streets and Highways, Bridges and Mass Transit.*

Corr, Arthur, "Capital Investment Planning," *Financial Planning,* August, 1980, pp. 12–15.

Council for Urban Development, *Coordinated Urban Economic Development,* Washington, DC, 1978.

Council of State Community Affairs Agencies, *State Financial Management Resource Guide,* Washington, DC, 1980.

Crinder, Robert A., *The Impact of Inflation on State and Local Government,* Urban and Regional Development Series, no. 5, Columbus, OH: Academy for Contemporary Problems, 1978.

Daily Bond Buyer, New York: Daily Bond Buyer, issued weekdays.

Dales, J.H., *Pollution, Property and Prices,* Toronto: University of Toronto Press, 1970.

Daniels, Belden, and Nancy Barbe, *New England Innovation: Paradigm for Reindustrialization,* Cambridge, MA: Counsel for Community Development, Inc., February 9, 1981.

Darst, David M., *The Handbook of the Bond and Money Markets,* 2d edition, New York: McGraw-Hill, 1981.

DeLucia, R.J., *An Evaluation of Marketable Effluent Permit Systems: Final Report,* Washington, DC: U.S. Environmental Protection Agency, GPO, 1974.

Denver Regional Council of Governments, *Capital Improve-*

ments Programming for Local Governments, Denver, 1975.

Devoy, Robert, and Harold Wise, The Capital Budget, Washington, DC: Council of State Planning Agencies, 1979.

Donnelly, Robert M., "Strategic Planning For Better Management," Managerial Planning, November/December, 1980, pp. 3–6, 41.

Dossani, Nazir, and Wilbur Steger, "Trends in U.S. Public Works Investment: Report on a New Study," National Tax Journal, Vol. 33, no. 2, June, 1980, pp. 97–110.

Douglas, Scott, "Determinants of Capital Budgets," paper presented at the 1977 national meeting of the American Society for Public Administration, Atlanta.

Downing, Paul B., User Charges and Service Fees, Tallahassee: Florida State University Press, 1980.

Dyl, Edward A., and Michael D. Joehnk, "Leasing As a Municipal Finance Alternative," Public Administration Review, November/December, 1978, pp. 557–62.

Economic Development Administration of the U.S. Department of Commerce, An Updated Evaluation of the EDA-Funded Industrial Parks—1968–74, Washington, DC: GPO, 1974.

Economics Research Associates, Impacts of Fees and Changes on Urban Recreation and Cultural Opportunities, San Francisco, 1979.

Edelman, Seth, The Career of the New York State Public Authority, Albany, NY: State University of New York at Albany, 1976.

Edmonds, Charles P., and William P. Hoyd, "Industrial Development Bond Financing," Financial Executive, April, 1981.

Executive Office of the President, Small Community and Rural Development Policy, Washington, DC: GPO, December 20, 1979.

_____, Office of Management and Budget, Managing Federal Assistance in the 1980s, Washington, DC: GPO, March, 1980.

_____, Office of Management and Budget, Special Analyses, Budget of the United States Government—Fiscal Year 1980, Washington, DC: GPO.

_____, Office of Management and Budget, Public Works as Countercyclical Assistance, Washington, DC: GPO, November, 1979.

_____, Office of Management and Budget, *Reorganization Study of Local Development Assistance Programs,* Washington, DC: GPO, December, 1978.

Feldman, Paul, "On the Optimal Use of Airports in Washington, D.C." *Socio-Economic Planning Science,* Vol. 1, 1967, pp. 21–39.

Feldstein, Martin, "Distributional Equity and the Optional Structure of Public Prices," *American Economic Review,* March, 1972, pp. 231–247.

Feuer, Albert, "Motor Fuel Tax Alternatives," *State Government,* 1978, pp. 11–17.

Finck, John A., and Howard Pike, *Infrastructure Rehabilitation: Where Do We Go from Here?* Albany, NY: New York State Department of Environmental Conservation, October 28, 1981.

Fischer, Philip J., et al., "Risk and Return in the Choice of Revenue Bond Financing," *Governmental Finance,* September, 1980, pp. 9–13.

Forbes, Ronald W., and John E. Peterson, *State Credit Assistance to Local Governments,* Boston: First Boston Corporation, 1978.

_____, *Cost of Credit Erosion in the Municipal Bond Market,* Chicago: Municipal Finance Officers Association, December, 1975.

Forbes, Ronald W., and Edward F. Renshaw, *State Bond Banks: Review of Present Developments and Needs for Bond Banks,* Chicago: Municipal Finance Officers Association, 1972.

Forty States Eye Motor Fuel Tax Boost, Washington, DC: Highway Users Federation, January, 1981.

Foster, Robert, *State Responses to the Adverse Impacts of Energy Development in Wyoming,* Report published by Laboratory of Architecture and Planning, MIT, Cambridge, MA, 1977.

Fujardo, Richard P., *Capital Budgeting: Guidelines and Procedures,* Research Reports in Public Policy, no. 9, Santa Barbara: Urban Economics Program, University of California at Santa Barbara, July, 1976.

Galambos, Eva, and Arthur Schreiber, "Pricing for Local Government: User Charges in Place of Taxes" in *Making Sense*

Out of Dallas, Washington, DC: National League of Cities, 1978.

Getzels, Judith, and Charles Thurow, *Local Capital Improvements and Development Management: Analyses and Case Studies,* Chicago: American Planning Association, June, 1980.

Gilmore, John S., and Mary K. Duff, *Boom Town Growth Management: A Case Study of Rock Springs–Green River, Wyoming,* Boulder: Westview Press, 1975.

Glaser, Sidney, "New Jersey's Limits on State and Local Spending: A Model for the Nation," *New Jersey Municipalities,* November, 1978.

Gold, Steven D., *How Restrictive Have Limitations on State Taxing and Spending Been?* Legislative Finance Paper no. 13, Denver: National Conference of State Legislatures, January, 1982.

_____, *Trends in the Magnitude and Character of State Debt,* Denver: National Conference of State Legislatures, 1981.

_____, *Property Tax Relief,* Lexington, MA: Lexington Books, D.C. Heath, 1979.

Gramlich, Edward M., and Harvey Galper, "State and Local Fiscal Behavior and Federal Grant Policy" in *Brookings Paper on Economic Activity, no. 1,* 1973.

Grossman, David A., *The Future of New York City's Capital Plan,* Washington, DC: The Urban Institute, 1979a.

_____, *Water Resources Priorities for the Northeast,* Washington, DC: The Consortium of Northeast Organizations, September, 1979b.

Grover, Kathryn, "Wholesale Water Pricing: A Cost-to-Serve Plan That Works," *American City and County,* Washington, D.C., November, 1980.

Hall, William K., "Changing Perspectives on the Capital Investment Process," *Long-Range Planning,* Vol. 12, February, 1979, pp. 37–40.

Hanke, S.H., "On the Current Crisis in Urban Water Supply," unpublished paper, Washington, DC: President's Council of Economic Advisors, 1981.

_____, *Options for Financing Water Development Projects,* Baltimore: Johns Hopkins University Press, 1976.

Hanke, S.H., and R.W. Wentworth, "On the Marginal Cost of Wastewater Services," *Land Economics*, November, 1981, pp. 196–210.

Harrington, John, *Packaging Housing Mortgage Loans: Strategies for California*, Washington, DC: Conference on Alternative State and Local Policies, 1979.

Hatry, Harry P., *Local Government Capital Infrastructure Planning: Current State-of-the-Art and State-of-Practice*, Washington, DC: The Urban Institute, 1980.

_____, *Maintaining the Existing Infrastructure*, Washington, DC: The Urban Institute, August 28, 1980.

Hendershott, Patric H., and Timothy W. Koch, *An Empirical Analysis of the Market for Tax-Exempt Securities*, New York: New York University, Graduate School of Business Administration, Center for the Study of Financial Institutions, 1977.

Henion, Lloyd, and Mark Ford, "Financing Highway Maintenance," *Journal of Contemporary Studies*, Vol. 4, no. 2, Spring, 1981, pp. 38–47.

Herships, David, and Leon Karvelis, *Effects of the Reagan Administration's Economic Recovery Plan on the Credit Standing of State and Local Governments*, New York: Merrill Lynch, Pierce, Fenner and Smith, Inc., December, 1981.

Higgins, James M., "Strategic Decision Making: An Organizational Behavioral Perspective," *Managerial Planning*, March/April, 1978, pp. 9–13.

Higgins, Tom, "Road Pricing: Should and Might It Happen?" *Transportation*, Vol. 8, June, 1979, pp. 1–17

Hoggan, D.H., *A Study of Feasibility of State Water Use Fees for Financing Water Development*, Logan, UT: College of Engineering, Utah State University, 1977.

Holland, Stuart, *The State as Entrepreneur*, London: Weidenfeld and Nicholson, 1972.

Holloway, Clark, and William King, "Evaluating Alternative Approaches to Strategic Planning," *Long-Range Planning*, Vol. 12, August, 1977, pp. 74–78.

Howard, Kenneth S., *Changing State Budgeting*, Lexington, KY: Council of State Governments, 1973.

Howell, James M., and Charles F. Stamm, *Urban Fiscal Stress:*

A Comparative Analysis of 66 U.S. Cities, Lexington, MA: Lexington Books, D.C. Heath, 1979.

Hoyle, Robert S., "Capital Budgeting Models and Planning: An Evolutionary Process," *Managerial Planning*, November/December, 1978, pp. 78–89.

Hubbell, Kenneth L., editor, *Fiscal Crisis in American Cities: The Federal Response*, Cambridge, MA: Ballinger Publishing Company, 1979.

Humphrey, Nancy, George E. Peterson and Peter Wilson, *The Future of Cincinnati Capital Plant, America's Urban Capital Stock Series, vol. 3*, Washington, DC: The Urban Institute, 1979.

Humphrey, Nancy, and Peter Wilson, "Capital Stock Condition in Twenty-Eight Cities," Washington, DC: The Urban Institute, unpublished, February, 1980.

Illinois Bond Watcher, Springfield, IL: Illinois Economic and Fiscal Commission, July, 1981.

Illinois General Assembly, Joint Committee on Long-Term Debt, *Report*, Springfield, IL, January 10, 1979.

Ingram, Robert, and Ron Copeland, "State Mandated Accounting, Auditing and Finance Practices and Municipal Bond Ratings," *Public Budgeting and Finance*, Spring, 1982, pp. 21–33.

Institute of Public Administration, *Financing Transit: Alternatives for Local Government: Executive Summary*, Washington, DC: GPO, 1980.

Irwin, David T., "Debt Management for State Government," *State Government*, 1979, pp. 9–17.

Jarrett, James E., and Jimmy E. Hicks, *The Bond Bank Innovation: Maine's Experience*, Lexington, KY: Council of State Governments, February, 1977.

Jones, Benjamin, *Restoring Municipal Credit: The New Jersey Qualified Loan Bond Program*, Lexington, KY: Council of State Governments, June, 1978.

Katzman, Martin T., "Measuring the Savings From Municipal Bond Banking," *Governmental Finance*, March, 1980, pp. 19–25.

_____, "Municipal Bond Banking: The Diffusion of a Public Finance Innovation," *National Tax Journal*, Vol. 33, no. 2, June, 1980, pp. 149–60.

Kaufman, Henry, *The Crowding of the Municipal Bond Market*, Salomon Brothers, New York City, August, 1981.

Kaus, Robert, "Jobs for Everyone," *Harpers*, October, 1982, pp. 11–17.

Kaynor, Edward R., "Uncertainty in Water Resources Planning in the Connecticut River Basin," *Journal of the American Water Resources Association*, December, 1978.

Keller, Charles W., "Pricing of Water," *Journal of the American Water Works Association*, Vol. 69, January, 1977, pp. 92–103.

Kidwell, David S., and Patric H. Hendershott, "The Impact of Advanced Refunding Bond Issues on State and Local Borrowing Costs," *National Tax Journal*, Vol. 3, no. 1, March 1978, pp. 93–100.

Kieschnick, Michael, *Taxes and Growth*, Washington, DC: Council of State Planning Agencies, 1981.

Kimball, Ralph C., "The Effect of a Taxable Bond Option on Borrowing Costs of State and Local Governments in the Northeast," *New England Economic Review*, March, 1978, pp. 21–31.

———, "Commercial Banks, Tax Avoidance, and the Market for State and Local Debt Since 1970," *New England Economic Review*, January/February, 1977a, pp. 21–32.

———, *Commercial Bank Demand and Municipal Bond Yields*, Boston: Federal Reserve Bank of Boston, 1977b.

Kish, T., "A Look at Self-Supporting Utilities," *Water Pollution Control Federation Journal*, Vol. 52, no. 11, November, 1980.

Klapper, Byron, "Municipal Commercial Paper," *Government Finance*, September, 1980, pp. 10–15.

Kolb, Klaus J., "Economic Development in Alaska: Responsibilities for Providing Infrastructure," unpublished paper, Kennedy School of Government, Harvard University, May, 1982.

Kopcke, Richard, and Ralph C. Kimball, "Investment Incentives for State and Local Governments," *New England Economic Review*, January/February, 1979, pp. 20–40.

Korbitz, William E., editor, *Urban Public Works Administration*, Washington, DC: International City Management Association, 1976.

Kunde, James E., and Daniel E. Berry, "Restructuring Local Economies through Negotiated Investment Strategies," forthcoming *Policy Studies Journal* symposium on "Public Policy for Communities in Economic Crisis."

Lake, Elizabeth R., et al., *Who Pays for Clean Water?: The Distribution of Water Pollution Control Costs*, Boulder: Westview Press, 1979.

Lamb, Robert B., and Stephen P. Rappaport, *Municipal Bonds: A Comprehensive Review of Tax-Exempt Securities and Public Finance*, New York: McGraw-Hill, 1980.

Langton, John, *Toll Road Financing: Description and Policy Implications*, Washington, DC: Association of American Railroads, 1981.

Lehan, Edward Anthony, "The Cast for Directly Marketed Small Denomination Bonds," *Governmental Finance*, Vol. 5, no. 9, September, 1980, pp. 3–7.

Leistritz, Larry F., and Steven H. Murdock, *The Socioeconomic Impact of Resource Development: Methods for Assessment*, Boulder: Westview Press, 1981.

Levatino-Donoghue, Adrienne, "Local Bonds for Housing," *Journal of Housing*, Vol. 36, no. 6, June, 1979, pp. 306–9.

Levine, Charles H., and Ira Rubin, editors, *Fiscal Stress and Public Policy*, Sage Publications, Beverly Hills: 1980.

Lindsay, Robert, editor, *The Nation's Capital Needs: Three Studies*, New York: New York Committee on Economic Development, 1979.

Littlefield, S.C., and G.F. Thomsen, "Aircraft Landing Fees: A Game Theory Approach," *The Bell Journal of Economics*, Vol. 8, 1977, pp. 201–232.

Litvak, Lawrence, and Belden Daniels, *Innovations in Development Finance*, Washington, DC: Council of State Planning Agencies, 1979.

Lu, Catherine, *State Responses to the Adverse Impacts of Energy Development in North Dakota*, Cambridge, MA: Laboratory of Architecture and Planning, MIT, 1977.

Lubick, Donald C., and Harvey Galper, "The Defects of Safe Harbor Leasing and What to do about Them," *Tax Notes*, March 15, 1982.

MacDonald, Keith, editor, *Northeast Urban Infrastructures: A*

Reader, Boston: Coalition of Northeast Municipalities, 1982.

McWatters, Ann Robertson, *Financing Capital Formation for Local Governments,* Research Report, 79-3. Berkeley: University of California, March, 1979, Institute of Governmental Studies.

Management Policies in Local Government Finance, Washington, DC: International City Management Association, 1981.

Maryland, Capital Debt Affordability Committee, *Report on Recommended Debt Authorization for Fiscal Year 1983, Submitted to the Governor and the General Assembly of Maryland,* Annapolis, August 1, 1981.

Matson, Morris C., "Capital Budgeting: Fiscal and Physical Planning," *Governmental Finance,* August, 1976, pp. 42–50, 58.

Matz, Deborah, *Trends in the Fiscal Condition of Cities: 1978–80,* Washington, DC: GPO, 1980.

Mentz, J. Robert, et al., "Leveraged Leasing and Tax-Exempt Financing of Major U.S. Projects," *Taxes,* August, 1980. pp. 553–60.

Mick, Susan R., *User Charges and Fees,* Washington, DC: The Urban Institute, 1981.

Miralia, Lauren M., "Municipal Bond Insurance Gaining in Acceptance," *ABA Banking Journal,* February, 1980, pp. 63, 65, 66.

Mitchell, William E., "Debt Refunding: The State and Local Government Sector," *Public Finance Quarterly,* Vol. 7, no. 3, July, 1979, pp. 323–37.

Moak, Lennox L., *Administration of Local Government Debt,* Chicago: Municipal Finance Officers Association of the United States and Canada, 1970.

Moak, Lennox L., and Kathryn W. Killian, *A Manual of Suggested Practice for the Preparation and Adoption of Capital Budgets by Local Governments,* Chicago: Municipal Finance Officers Association of the United States and Canada, 1964.

Monaco. Lynne, *State Responses to the Adverse Impacts of Energy Development in Colorado,* Cambridge, MA: Laboratory of Architecture and Planning, MIT, 1977.

Moody's Bond Survey, New York: Moody's Investors Service, Inc. weekly. *Moody's Municipal and Government Bonds: News*

Report, New York: Moody's Investors Service, Inc. Issued every Tuesday and Friday.

Moody's Municipal and Government Manual—American and Foreign, New York: Moody's Investors Service, Inc. Since 1955, annually.

Morgan Guaranty Survey, *Fiscal Stress for States and Localities,* New York: Morgan Guaranty Trust Co., November, 1981.

Mumy, Gene E., "Issue: Costs and Competition in the Tax-Exempt Bond Market," *National Tax Journal,* Vol. 3, no. 1, March, 1978, pp. 81–91.

Municipal Finance Officers Association of the United States and Canada, *A Guidebook to Improved Financial Management in Smaller Municipalities,* Chicago, August, 1978.

_____, *Costs Involved in Marketing State/Local Bonds,* Chicago, 1976.

Mushkin, Selma J., "The Case for User Fees," *Taxes and Spending,* April, 1979, pp. 49–62.

_____, editor, *Public Prices for Public Products,* Washington, DC: The Urban Institute, 1972.

_____, "Prices as an Alternative to Reorganization," paper presented at Conference on Government Reorganization, Woodrow Wilson School, September, 1977.

Mushkin, Selma J., and Charles L. Vehorn, "User Fees and Charges," *Governmental Finance,* Vol. 6, November, 1977, pp. 61–75.

Mussa, Michael L., and Roger C. Kormendi, *The Taxation of Municipal Bonds: An Economic Appraisal,* Washington, DC: American Enterprise Institute for Public Policy Research, 1979.

National Association of Counties, *Bridging the Revenue Gap,* Washington, DC, 1980.

National Conference of State Legislatures, *Guidelines for Single-Family Tax-Exempt Mortgage Revenue Bonds,* Denver, 1980, p. 30.

National Governors' Association, *Federal Roadblocks to Efficient State Government,* Washington, DC, 1977.

National Governors' Association for Policy Research, *Bypassing the States: Wrong Turn on Urban Aid,* Washington, DC, November, 1979.

Naylor, Thomas H., "Organizing for Strategic Planning," *Managerial Planning,* July/August, 1979, pp. 3–9, 17.

Naylor, Thomas H. and Kristin Neva, "The Design of a Strategic Planning Process," *Managerial Planning,* January/February, 1980, pp. 3–7.

_____, "The Planning Audit," *Managerial Planning,* September/October, 1979, pp. 31–37.

Neuner, Edward, Dean Dopp, and Fred Sebold, "User Charges versus Taxation as a Means of Funding a Water Supply System," *Journal of the American Water Works Association,* Vol. 69, 1977, pp. 256–281.

Newsweek, "The Decaying of America," August 2, 1982, p. 25.

Northeast-Midwest Institute, *Urban Water Supply and Sewer Needs in the Midwest,* Washington, DC, December, 1980.

Oates, Wallace E., "The Use of Local Zoning Ordinances to Regulate Population Flows and the Quality of Local Services" in *Essays in Labor Market Analysis,* edited by Orly Ashenfelter and Wallace E. Oates, New York: John Wiley and Sons, 1977.

O'Hare, Michael, "Not on My Block You Don't—Facility Siting and the Strategic Importance of Compensation," *Public Policy,* Vol. 25, no. 4, 1977, pp. 78–93.

O'Hare, Michael, and Debra Sanderson, "Fair Compensation and the Boomtown Problem," *Urban Law Annual,* Vol. 14, 1978, pp. 29–37.

Osteryoung, Jerome S., "State General Obligation Bond Credit Ratings," *Growth and Change,* Vol. 9, no. 3, July, 1978, pp. 95–103.

Pagano, Michael, and Richard J. Moore, "Emerging Issues in Financing Basic Infrastructure," unpublished paper, September, 1981.

Pascal, Anthony, *User Charges, Contracting Out, and Privitization in an Era of Fiscal Retrenchment,* P-6471, Santa Monica, CA: The Rand Corporation, April, 1980.

Peterson, George E., *An Examination of State and Local Governments' Capital Demand, Alternative Means of Financing 'Public' Capital Outlays and the Impact on Tax-Exempt*

Credit Markets, Washington, DC: The Urban Institute, n.d., pp. 65–69.

_____, *Tax-Exempt Financing of Housing Investment,* Washington, DC: The Urban Institute, 1979.

Peterson, George E., and Mary John Miller, *Financing Infrastructure Renewal: Policy Options,* Washington, D.C., Urban Consortium, December, 1981.

Peterson, John E., "Has the Municipal Bond Market Undergone Fundamental Change?" unpublished paper, May, 1982.

_____, "Current Research in State and Local Government Debt Policy and Management, parts 1, 2, *Government Finance,* March/June, 1979a, pp. 45–48: November, 1978, pp. 33–35.

_____, *State Roles in Local Government Financial Management: A Comparative Analysis,* Washington, DC: Government Finance Research Center, 1979b.

_____, *State and Local Government Finance and Financial Management: Compendium of Current Research,* Chicago: Municipal Finance Officers Association of the United States and Canada, 1978, p. 27.

_____, *Watching and Counting: A Survey of State Assistance to and Supervision of Local Debt and Financial Administration,* Denver, National Conference of State Legislatures and the Municipal Finance Officers Association, 1977.

_____, *Changing Conditions in the Market for State and Local Government Debt,* Washington, DC: GPO, 1976.

_____, *The Rating Game,* New York City, Twentieth Century Fund, 1974.

"Preserving U.S. Roads: A Rough Time Ahead," *Journal of American Insurance,* Winter, 1977–78, pp. 16–20.

Public Securities Association, *Fundamentals of Municipal Bonds,* New York, 1981.

"Reform of the Municipal Bond Market: Alternatives to Tax-Exempt Financing," *Columbia Journal of Law and Social Problems,* Vol. 15, no. 3, Fall, 1979, pp. 233–75.

Reilly, James F., *Municipal Credit Evaluation and Bond Ratings Diagnosis, Prognosis and Prescription for Change,* Berkeley, CA: Institute for Local Government, 1967.

Robinson, Donald J., et al., *Municipal Bonds 1981—A Course Handbook,* New York: Practicing Law Institute, 1981.

Rocha, Luis M., "Post-Development Costs in Rural Communities in Alaska," unpublished paper, Kennedy School of Government, Harvard University, May, 1982.

Rothschild, L.F., Unterberg, Towbin, Municipal Research Department, *The Fifty States: Will Budget Problems Continue?* New York, January 11, 1982.

Rumowicz, Madelyn, "In New Jersey: Capital Budgeting and Planning Process," *State Government,* Spring, 1980, pp. 99–102.

Saffran, James S., "Proposition 13: Effect Upon the Bond Market," *Western City,* February, 1979.

Sanderson, Debra, *State Responses to the Adverse Impacts of Energy Development in Texas,* Cambridge, MA: Laboratory of Architecture and Planning, MIT, 1977.

Schellenbach, Peter W., and James S. Weber, "Leasing: An Alternative Approach to Providing Governmental Services and Facilities," *Government Finance,* November, 1978, pp. 23–27.

Schilling, Paul R., "Wisconsin Municipal Debt Finance: An Outlook for the Eighties," *Marquette Law Review,* Vol. 63, no. 4, Summer, 1980, p. 539–92.

Schmidt, Richard, "Strategic Planning: Off-Limits for Financial Managers?" *Management Review,* June, 1979, pp. 71–77.

Schneider, Mark, and David Swinton, "Policy Analysis in State and Local Government," *Public Administration Review,* January/February, 1979, pp. 12–17.

Schnell, John F., and Richard S. Krannich, *Social and Economic Impacts of Energy Development Projects: A Working Bibliography,* Monticello, IL: Council of Planning Librarians, 1977.

Schramm, Gunter, *The Value of Time in Environmental Decision Processes,* Ann Arbor, MI: The University of Michigan, November, 1979.

Schwartz, Gail Garfield, and Pat Choate, *Being Number One: Rebuilding the U.S. Economy,* Lexington, MA: Lexington Books, D.C. Heath, 1980.

Schwertz, Eddie L., Jr., *The Local Growth Management Guidebook,* Washington, DC: The Southern Growth Policies Board, 1979.

Shaul, Marnie S., "The Determinants of City Borrowing," Ph.D. dissertation, Ohio State University, 1980, pp. 135–37.

————, "Capital Financing Options for Local Government," unpublished paper, National Urban Policy Roundtable, 1980.

Sheeran, Burke F., *Management Essentials for Public Works Administrators,* Chicago: American Public Works Association, 1976.

Shepard, Kevin, and Haynes C. Goddard, "New Approaches to Capital Planning and Financing," National Urban Policy Roundtable discussion paper, n.d.

Shoup, Donald C., "Financing Public Investment by Deferred Special Assessment," *National Tax Journal,* Vol. 33, no. 4, December, 1980, pp. 413–29.

Shubnell, Larry, and Bill Cobb, "Creative Capital Financing: A Primer for State and Local Governments," *Resources in Review,* May, 1982, pp. 7–11.

Silber, William L., *Municipal Revenue Bond Costs and Bank Underwriting: A Survey of the Evidence,* New York: New York University, Graduate School of Business Administration, Salomon Brothers Center for the Study of Financial Institutions, 1980.

Small Cities Financial Management Project, *A Debt Management Handbook for Small Cities and Other Governmental Units,* Chicago: Municipal Finance Officers Association of the United States and Canada, 1978.

Smith, Fred L., *Alternatives to Motor Fuel Taxation—Weight-Mileage Taxes,* Washington, DC: Association of American Railroads, 1980.

Smith, Wade S., *The Appraisal of Municipal Credit Risk,* New York: Moody's Investor Service, Inc., 1979.

Solano, Paul, and Steven Hoffman, "Municipal Bond Banking: A Comment," *National Tax Journal,* March, 1982, pp. 64–79.

Stamm, Charles F., and James M. Howell, "Urban Fiscal Problems: A Comparative Analysis of 66 U.S. Cities," *Taxing and Spending,* Fall, 1980, pp. 41–58.

Stanfield, Rochelle, "Building Streets and Sewers Is Easy—It's Keeping Them Up That's the Trick," *National Journal,* May 24, 1980, pp. 1141–1145.

Stanley, David T., *Cities in Trouble*, Columbus, OH, Academy for Contemporary Problems, 1976.

Steiss, Walter Alan, *Local Government Finance: Capital Facilities Planning and Debt Administration*, Lexington, MA: D.C. Heath, 1975.

Sternlieb, George, *Housing Development and Municipal Costs*, New Brunswick, NJ: The Rutgers Center for Urban Policy Research, 1974.

Susskind, Lawrence, and Michael O'Hare, *Managing the Social and Economic Impacts of Energy Development*, Cambridge, MA: Laboratory of Architecture and Planning, MIT, 1977.

_____, Task Force on Municipal Bond Credit Rating, *The Rating Game*, New York: Twentieth Century Fund, 1974.

U.S. Bureau of the Census, *1977 Census of Governments*, Washington, DC: GPO, 1980. In 7 volumes.

_____, *Statistical Abstract of the United States: 1980*, Washington, DC: GPO, 1979.

_____, *State Government Finances*, Washington, DC: GPO, 1964.

_____, *Local Government Finances in Selected Metropolitan Areas and Large Counties*, Washington, DC: GPO, 1964.

U.S. Conference of Mayors, *Transit Financing: An Overview of the National Transit Financing Picture in Terms of Federal and State Funding Levels, Fare Structures and Local Revenue Sources*, Washington, DC: GPO, October, 1980 (DOT-P-30-80-34).

U.S. Congress, House Subcommittee on the City of the Committee on Banking, Finance, and Urban Affairs, *City Need and the Responsiveness of Federal Grant Programs*, Peggy L. Cuciti Subcommittee Print, Washington, DC: GPO, 1978.

_____, Committee on Public Works, *A National Public Works Investment Policy*, Washington, DC: GPO, December, 1974.

_____, Joint Economic Committee, *Chaos in the Municipal Bond Market*, Washington, DC: GPO, September 28, 1981.

_____, Joint Economic Committee, *Trends in the Fiscal Conditions of Cities, 1979–1981*, Washington, DC: GPO, May, 1981.

_____, Joint Economic Committee, *Public Works as a Countercyclical Tool*, Washington, DC: GPO, 1980.

_____, Joint Economic Committee, Subcommittee on Economic Growth and Stabilization, *Deteriorating Infrastructure in Urban and Rural Areas: Hearing, August 30, 1979*, Washington, DC: GPO, 1979.

_____, Senate Committee on Public Works, United States Senate, 93rd Congress, 2d session, *Construction Delays and Unemployment*, Washington, DC: GPO, 1974.

_____, *Senate Committee on Environment and Public Works, Hearings on the Inland Energy Development Impact Assistance Act of 1977 (S 1493)*, Washington, DC: GPO, 2 parts, August 2, 27, 1977, and May 10, June 19, 1978.

_____, Senate Committee on Governmental Affairs, Subcommittee on Intergovernmental Relations, *Intergovernmental Fiscal Impact of Mortgage Revenue Bonds: Hearing, July 18, 1978*, Washington, DC: GPO, 1979.

U.S. Department of Agriculture, *Rural Development Progress, January 1977–June 1979*, Washington, DC: GPO, 1979.

_____, *Social and Economic Trends in Rural America*, Washington, DC: GPO, October, 1979.

U.S. Department of Commerce, *Establishment of a National Development Bank*, unpublished paper, National Public Advisory Committee on Regional Economic Development, Washington, DC: GPO.

_____, *Governmental Finances in 1979–80*, Washington, DC: GPO, 1981.

_____, *A Study of Public Works Investment in the United States*, Washington, DC; GPO, 1980.

U.S. Department of Defense, *Boom Town Annotated Bibliography*, Washington, DC: The Pentagon, 1981a.

_____, *Boom Town Business Opportunities and Management Development*, Washington, DC: The Pentagon, 1981b.

_____, Office of Economic Adjustment, *Base Closures: Are the Economic Impact Predictions Realistic? An Analysis of the Post-Closure Economic Impact of Military Installations on Local Communities*, Washington, DC: The Pentagon, 1979.

_____, *Communities in Transition, Community Response to Reduced Defense Activity*, Washington, DC: The Pentagon, 1978.

U.S. Department of Housing and Urban Development, *Streamlining Land Use Regulation: What Local Public Officials Should Know, Washington, DC: GPO, 1980.

_____, The President's National Urban Policy Report, Washington, DC: GPO, 1980.

_____, Causes and Consequences of Delay in Implementing the Community Development Block Grant Program, Washington, DC: GPO, June, 1980.

_____, Advance Project Planning for Public Works: A Systematic Approach, Washington, DC: GPO, 1979.

U.S. Department of Transportation, The Status of the Nation's Highways: Condition and Performance, Washington, DC: GPO, January, 1981.

_____, Federal Highway Administration, 1981 Federal Highway Legislation: Program and Revenue Options, Washington, DC: GPO, June, 1980.

_____, Draft Transportation Agenda for the 1980s: The Issues, Washington, DC: GPO, March, 1980.

_____, Financing Transit: Alternatives for Local Government, Washington, DC: GPO, July, 1979.

U.S. Environmental Protection Agency, 1990: Preliminary Draft Strategy for Municipal Waste Water Treatment, Washington, DC: GPO, 1981.

_____, Clean Water: Fact Sheet, Washington, DC: GPO, April, 1980.

_____, The Cost of Clean Air and Water: A Report to the Congress, Washington, DC: GPO, August, 1979.

U.S. General Accounting Office, Effective State and Local Capital Budgeting Practices Can Help Arrest the Nation's Deteriorating Infrastructure, Washington, DC: GPO, June, 1982.

_____, Better Targeting of Federal Funds Needed to Eliminate Unsafe Bridges, Washington, DC: GPO, August, 1981.

_____, More Can Be Done to Insure that Industrial Parks Create New Jobs, Washington, DC: GPO, December 2, 1980.

_____, Foresighted Planning and Budgeting Needed for Public Buildings Program, Washington, DC: GPO, September 9, 1980.

_____, Perspectives on Intergovernmental Policy and Fiscal Relations, Washington, DC: GPO, June 28, 1978.

_____, *Federally Assisted Areawide Planning: Need to Simplify Policies and Practices*, Washington, DC: GPO, March, 1977.

_____, *Long-Range Analysis Activities in Seven Federal Agencies*, Washington, DC: GPO, December, 1976.

Vance, Mary A., *Municipal Bonds: A Bibliography*, Monticello, IL: Vance Bibliographies, August, 1981.

Vaughan, Roger J., "Federal Tax Policy and State and Local Fiscal Conditions" in *The Urban Impacts of Federal Policies*, Norman Glickman, ed., Baltimore: Johns Hopkins University Press, 1980.

_____, *Inflation and Unemployment: Surviving the 1980s*, Washington, DC: Council of State Planning Agencies, 1980a.

_____, "Countercyclical Public Works: A Rational Alternative," testimony given before the Joint Economic Committee of the United States Congress, June 17, 1980b, Washington, DC.

_____, *State Taxation and Economic Development*, Washington, DC: Council of State Planning Agencies, 1979.

_____, *Public Works as a Countercyclical Device: A Review of the Issues*, Santa Monica, CA: The Rand Corporation, July, 1976.

Vernez, Georges, and Roger J. Vaughan, *Assessment of Countercyclical Public Works and Public Service Employment Programs*, Santa Monica, CA: The Rand Corporation, 1978.

_____, and Robert K. Yin, *Federal Activities in Urban Economic Development*, Santa Monica, CA: The Rand Corporation, April, 1979.

_____, and Burke Burright, and Sinclair Coleman, *Regional Cycles and Employment Effects of Public Works Investments*, Santa Monica, CA: The Rand Corporation, 1977.

Viscount, Francis, *Municipal Bonds: The Need to Regulate*, Washington, DC: National League of Cities, May, 1982.

Vogt, John A., *Capital Improvements Programming: A Handbook for Local Officials*, Chapel Hill: Institute of Government, University of North Carolina, 1977.

Wacht, Richard F., *A New Approach to Capital Budgeting for City and County Governments*, Research Monograph no. 87, Atlanta: College of Business Administration, Georgia State University, 1980.

Wallace, Holly, "Infrastructured: Maintain It Now or Pay the Price Tomorrow" in *City Economic Development*, Washington, DC: National League of Cities, May 12, 1980.

Walsh, Ann Marie Hauck, *The Public's Business: The Policies and Practices of Government Corporations*, Cambridge, MA: MIT Press, 1978.

West, Stanley, *Opportunities for Company-Community Cooperation in Mitigating Energy Facility Impacts*, Cambridge, MA: Laboratory of Architecture and Planning, MIT, 1977.

White, Anthony G., *Municipal Bonding and Taxation*, New York: Garland, 1979.

White House Conference on Balanced National Growth, proceedings, Washington, DC: GPO, 1978.

White, Michael J., "Capital Budgeting," in *State and Local Government Finance and Financial Management*, ed. John E. Petersen et al., Chicago: Municipal Finance Officers Association, 1978.

White, Michael J., and Scott Douglas, "An Interpretation of Capital Programming as a Political Process in No-Growth Municipalities," paper presented at the 1975 annual meeting of the American Political Science Association, San Francisco.

White, Sharon S., *Municipal Bond Financing of Solar Energy Facilities*, Washington, DC: GPO, 1980.

Wildavsky, Aaron, *Budgeting: A Comparative Theory of Budgetary Processes*, Boston: Little, Brown and Company, 1975.

Wilson, Peter, *The Future of Dallas' Capital Plant*, Washington, DC: The Urban Institute, 1980.

Wolman, Harold, and Barbara Davis, *Local Government Strategies to Cope with Fiscal Pressure*, Washington, DC: The Urban Institute, 1980.

Wolman, Harold, and George Reigeluth, *Financial Urban Public Transportation: The U.S. and Europe*, New Brunswick, NJ: Transaction Books, 1980.

Worsham, John P., Jr., *Tax-Exempt Mortgage Bonds: Revenue Aids to Homeownership*, Monticello, IL: Vance Bibliographies, November, 1980.

Zeller, Martin, *The Management of Mineral Revenues in the Western Energy-Producing States*, unpublished paper prepared for the Council of State Planning Agencies, Washington, DC, 1982.

Zimmerman, Joseph, *Reassignment of Functional Responsibility,* U.S. Advisory Commission on Intergovernmental Relations, Washington, DC: GPO, July, 1976.

Index

ments, demographic changes and, 34–35

Education debt financing, 92; local government responsibility, 14; new skills and, 13, 56; state and local expenditure for, 25, 38; on state responsibilities, 182

Education facilities, investment in, 14–15

Efficiency, definition of, 42

Electric utilities, sale to private companies, 162

Electricity generation, in water projects, 179; peak-load pricing for, 54; privately provided, 39

Electricity hook-up, subsidies, 67

Electricity, privately provided, 39

Energy conservation, user fees and, 49

Energy costs, local government response to, 37

Engineering standards, insufficient, 172; to plan public investments, 40, 42; redesigning, 5

Environmental issues, public sector responsibility, 14

Environmental Protection Agency, waste water treatment financed by, 18

Equipment, depreciation of, 146–147; lease limits, 149

ERTA, 148; and safe harbor leases, 149, 150, 151

Evaluation, financing of, 138; of policies and programs, for states and localities, 182

Externalities, definition of, 43

Exxon, 123

Federal aid, historically, 18; reduction in, 3, 8, 30, 171, 185

Federal borrowing, as share of net savings, 95

Federal business tax incentives, state and local governments use, 142–170

Federal capital budget, 175

Federal debt, 75

Federal development bank, low cost loans from, 175

Federal fiscal assistance, to localities, 19, 38, 175–176

Federal grants, change investment priorities, 18

Federal Interstate Highway Trust Fund, 61–62

Federal loans, to states and localities, 175–176

Federal mineral leasing royalty payments, in North Dakota, 132; in Wyoming, 132

Federal planning and management, 175, 181–182

Federal policies, and infrastructure problems, 174; and state and local investment strategies, 174–182

Federal priorities, versus local priorities, 21–22

Federal public assistance, state matching for, 25

Federal role, in boom town, 136; in financing development projects, 136–137; in funding R & D, 181

Federal subsidies, design of, 41–42

Federal transfers, to state and local governments, 24

Financial intermediation, state role, 99–103

Fire department equipment, and safe harbor leasing, 150

Fire protection, municipal responsibility, 38

Funds, targeted to impact areas, 44–45

Safe harbor lease, 149, 151–155, 167

Sale of public facilities, to private firms, 142, 162–167

Sale/leaseback, 155

Sales and property tax, increment financing of, 99

Sales tax, advance planning for use, 44; in boom towns, 126; increment financing of, 99; earmarked, 70; and limited obligation bonds, 139; regressive, 57; revenue for state governments, 24

San Francisco, investments cut, 17

Sanitation, municipal responsibility, 38

School district funds, limited, 30; in Montana, 133–134; in North Dakota, 130, 132; in Wyoming, 132

Schools, in boom town, 110

Section "38" property, 148

Service charges, 24

Service contract, 142, 150, 155–162, 168

Service delivery, innovation in, 66

Service financing, indirect way of reducing costs, 172

Severance tax, 124, 125, 127–130, 139

Sewage hook-up, subsidies, 67

Sewer, debt financing of, 92; local expenditure for, 25; marginal costs of, 50; refinancing of in New York, 168; sale/leaseback agreement, 167; state spending for, 22

Short-term debt, increase in, 83–91

Short-term financing, high interest rates cause, 75

Social programs, expenditures on, 18

Solid waste management, 115

St. Petersburg, capital spending cut, 17; deferred maintenance, 4

State actions, reduce cost of issuing bonds, 74

State and local debt, 75–92, 93

State and local government, aid to private for-profit firms, 38; borrowing subsidies, 94; compete with federal government in securities, 96; diversify fiscal bases, 23; responsibility, history of, 38–39

State and local guidelines, who does what, 40

State and local officials, powerless, 3

State and local policy options, within federal actions, 174

State and local public services, demand for, 34

State and local tax revenues, recession shrunk, 3

State assistance, to local governments, 7–8, 182, 184–185

State assumption, of local responsibilities, 8

State bond banks, 7, 74

State bond rating, 104

State borrowing, constitutional limits on, 83, 84–85

State capital budgets, 182

State debt, growth of, 75; limit on GOBs, 83

State economies, future of, 35

State fiscal capacity, public investments and, 30, 34

State funds, allocation of by need, 45

State government, boom town responsibilities for, 136–137; expenditures, 25–28; increased responsibility for capital investments, 174

Taxes levied by, special assessment districts, 99
Technological changes, future demands for, 15, 34
Telecommunications systems, service contracts for, 157
Terminals, 178, 183
Texas Water Development Fund, 102–103
Texas, assistance to localities, 134–135; loan program for water, 102–103; resource revenues finance other activities, 139; wants more energy development, 137
Tires, tax on, 61
Toll bridges, 40, 66, 162–163
Toll revenues, user fees for highways, 62, 180
Topsoil, poor management of, 35
Traffic courts, self financed, 63
Traffic violation fines, increase in, 62
Transfer of responsibility, from private to public, 38–39
Transit fares increase, 39
Transit lines, sale to private companies, 162
Transit service, increases nearby property value, 71
Transit system, peak-load pricing for, 51
Transit, financing solutions, 66; highway construction funds used for, 22; and low-income households, 65, 66; maintenance of, 41, 42; state spending for, 22; subsidies for, 41, 65–66; user fee for, 183
Transportation, capital investment in, 13; debt financing of, 92; delegation to state government, 174; equipment depreciation of, 146; private

ownership of, 70; public investments in, 35
Trucks, service contracts and, 157
Trust fund revenues in mineral-rich states, 139–140
Tuition increases, and extra help for poor, 63

Underground economy, 30
Universities, user fees for, 63–65
University research, funding of, 63–65
Urban Development Action Grant Program, 176
Urban fiscal crises, causes of, 19
Urban Mass Transit Administration, 41
Urban streets, public investment in, 13–14
User fees, 7, 47–73; amount collected, 49; beneficiaries pay, 59; cash grants, 56–57; conservation, 49; construction and operation, 172; defined, 9, 47; disadvantages of, 43, 57; economic development, 57–58; effect of, 72, 172; fairness of, 49; for inland water transportation, 175; limits to, 54–57, 72; LOB's and, 42; low-income households, 59, 173; medical research facilities, 63–65; peak-load pricing and, 59; principles of, 59; property taxes contrasted with, 49; public works spending increase, 8; reduces budget flexibility, 58; resource conservation tool, 54; state assistance for, 183, 185; state capitol investment programs,

STUDIES IN DEVELOPMENT POLICY

1. *State Taxation and Economic Development* by Roger J. Vaughan
2. *Economic Development: Challenge of the 1980s* by Neal Peirce, Jerry Hagstrom, and Carol Steinbach
3. *Innovations in Development Finance* by Lawrence Litvak and Belden Daniels
4. *The Working Poor: Towards a State Agenda* by David M. Gordon
5. *Inflation and Unemployment* by Roger J. Vaughan
6. *Democratizing the Development Process* by Neal Peirce, Jerry Hagstrom, and Carol Steinbach
7. *Venture Capital and Urban Development* by Michael Kieschnick
8. *Development Politics: Private Development and the Public Interest* by Robert Hollister and Tunney Lee
9. *The Capital Budget* by Robert DeVoy and Harold Wise
10. *Banking and Small Business* by Derek Hansen
11. *Taxes and Growth: Business Incentives and Economic Development* by Michael Kieschnick
12. *Pension Funds and Economic Renewal* by Lawrence Litvak
13. *The Road to 1984: Beyond Supply Side Economics* by Roger J. Vaughan

POLICY PAPERS

1. *Economic Renewal: A Guide for the Perplexed*
2. *The Employment and Training System: What's Wrong With It and How To Fix It*
3. *State Regulation and Economic Development*
4. *Industrial Policy*
5. *Financing Entrepreneurship*
6. *State and Local Investment Strategies*
7. *Monetary Policy*
8. *Social Security: Why It's In Trouble and What Can Be Done About It*
9. *Pension Funds and the Housing Problem*

STUDIES IN RENEWABLE RESOURCE POLICY

1. *State Conservation and Solar Energy Tax Programs* by Leonard Rodberg and Meg Schachter
2. *Environmental Quality and Economic Growth* by Robert Hamrin

America in Ruins: Beyond the Public Works Pork Barrel by Pat Choate and Susan Walter

The Game Plan: Governance With Foresight by John B. Olsen and Douglas C. Eadie